U0288820

2018 第十一届IAI全球设计奖年鉴

IAI DESIGN AWARD YEARBOOK

何昌成 编

上海书画出版社

IAI简介 ABOUT IAI

IAI全球设计奖，简称“IAI”，是亚太地区乃至国际最具影响力和美誉度的设计大奖之一，作为亚洲最具创新力的大奖，IAI评审团根据出色的设计作品评定出获奖名次和奖项荣誉，IAI设计奖已经被视为世界范围内新的拥有高识别度和含金量的国际设计大奖。

IAI设全球计奖始自2006年举办的亚太室内设计双年大奖赛，期间经历三次奖项名称更迭，并于2014年更名为全设计领域的设计奖为“IAI设计奖”，被中央电视台等国内外权威媒体赞誉为“设计奥斯卡”。2019年，IAI设计奖再次蜕变，将IAI设计奖划分为三大子奖项暨“IAI智造奖”“IAI建筑奖”“IAI室内奖”，完善并提升了IAI设计奖的专业高度，进一步奠定了IAI奖项的权威性和美誉度。

IAI全球设计奖颁发给在创新和可持续设计上做出杰出贡献的设计师，院校和设计机构等；每年IAI都会给获奖者出版年鉴，并在全球发行，寻求商业合作； IAI设计奖包含室内设计、建筑设计、工业产品设计、传播设计和产品概念设计。APDF在亚洲设立IAI年度大奖的目的是将设计创意和商业牵线搭桥。

IAI Design Awardabbreviation “IAI” ,it is one of the most international influence and reputation design awards in the Asia Pacific region and even the world.The IAI Design Awards are hosted annually by the APDF (Asia Pacific Designers Federation). It has been successfully held for 11 years and has been praised as “Designing Oscars” by CCTV and other domestic and international authoritative media. As the most innovative award in Asia, the IAI juries evaluate awards based on outstanding design work. The IAI Design Award has been recognized as the world's new international design award with high recognition and gold content.

In 2019, the IAI Design Award was once again transformed. The IAI Design Award was divided into three sub- awards and IAI Interior Awards (IAI+I), IAI Architectural Award (IAI+A), and IAI Intelligence Manufacturing Award (IAI+IM). It also enhanced the professionalism of the IAI Design Award and further established the authority and reputation of the IAI Awards.

The IAI Design Awards are presented to designers, institutions and design agencies who have made outstanding contributions to innovative design and sustainable design. Each year, IAI publishes yearbooks for winners and distributes them globally for business collaboration; the IAI Design Awards include Interior design, architectural design, industrial product design, communication design and product concept design. APDF sets up the IAI Annual Awards in Asia to match design ideas and business.

Content

何昌成 Oskar Ho

IAI设计奖发起人

APDF创始主席

IAI Design Award Promoter

Founding President of Asia Pacific Designers Federation

序言

2018第十一届IAI设计奖(以下简称“IAI”)与上届相比出现了如下几个新特点，一是获奖作品类别比往届丰富；二是作品国际化程度比往届高；三是获奖作品含金量高；四是新锐佳作涌现；五是工业设计产品尤其是智能产品数量和质量都有了显著提升。本届大赛作品除了来自中国包括香港和台湾地区外，还有美国、德国、意大利、日本、新加坡、瑞士、希腊、智利、以色列、马来西亚、乌克兰等国家和地区的1376件优秀作品参赛，195件作品入围角逐IAI各项设计大奖，最终经过三轮评审，评出了42项大奖，其中包括室内、建筑、工业产品设计三大类别的大奖和杰出设计大奖、杰出设计机构、评审团特别奖、最佳绿色环保大奖、最佳材料创新大奖、最佳人文关怀大奖、设计之星奖。更令人欣喜的是“IAI大奖”获得者还涌现了一大批新面孔。

在室内空间类别，日本的设计师大沽·日比野拓的三个幼儿教育空间作品不仅体现了设计美学，而且在其充满了童趣与益智的空间中融入了人文关怀和人性的理念；来自新加坡公司Ministry of Design设计的《竞赛机器人实验室》摘得了IAI最佳设计大奖— 展示空间的大奖；广州设计师黄永才的《洞穴俱乐部》利用洞穴的概念展现人与空间、空间与空间的关系，既独立又相互渗透；苏州叙品空间的《水云间茶会所》以现代中式风格展示了一个素雅、别致的茶文化空间；张建武和蔡天宝的商业空间作品《层迭》则巧妙地利用有限的空间结构进行巧妙的布局，并通过室内的装置和“黑与黄”的经典而充满时尚感的色彩语言形成强烈的视觉冲击；姜晓林的《阳朔花梦间酒店》从建筑到景观到室内再到软装设计，将自然景观融入室内空间，令空间达到如梦般的感觉，流露出一种素雅清新的东方美学独特气质。此外，拿云室内设计有限公司（中国大陆）的《零宠物俱乐部》，美国何丹尼的《一天》，欧阳昆仑、何魏冰（中国大陆）的《文成堂》，创空间集团（中国台湾）的《余乐园》，洪逸安（中国台湾）的《我的秘密花园》，钱银玲（中国大陆）的《泉上海鲜粥》都展现了不同的空间美学和特质。

在建筑设计类别，世博会以色列馆总设计师渡堂海的《张家界玻璃天桥》采用了隐形的设计，使天桥自然而然地消失在白云中，远看玻璃桥薄如蝉翼，达到了与自然环境的和谐，该大桥创造了最长玻璃桥面人行桥、世界最陡溜索、世界首座大张开量空间索面悬索桥等十项世界之最。建筑师魏娜的《小溪家》，通过改造福建闽东一个村子里小溪崖边一栋被遗弃的老房子和羊圈，将它们变为一个充满温情的“家”一样的民宿，令人感到鼓舞和温暖；凌子达的建筑作品《武汉东源售楼处》以其一贯的充满雕塑感和线条感、富于大胆创新的空间建构思维，展示了鲜明的个人设计风格；瑞士建筑师德科乌维·蒙许的《华翔堂社区中心》则展现了一个被购物中心和办公大楼中所包围的社区小教堂，通过改造后成为当地一抹亮丽的地标建筑；而《SLT办公室》以其舒适且温暖的色调、亲和的环保材料，以及绿色的自然元素令人印象深刻。李硕的民宿建筑《倾城客栈》则尊重当地传统文化，将自然生态与建筑群落和谐地融为一体，颇让人惊叹中国传统文化的无穷魅力。

在工业设计类别，意大利设计师戴维德·蒙塔纳罗的《鳞灯系列》灵感来自鱼鳞，以钢片的金属质感制作出鳞灯的不同造型和创意，展示了现代工具工艺与科技的完美融合。来自希腊的设计师尼科·卡帕的概念产品《高迪系统》体现了独特的创新性和人性化及环境友好理念；此外，易成海的《活字茶盘》、深圳前海帕拓逊公司设计研究的《噪音发生器》、中国台湾实践大学的学生瞿伟民的《木语》、刘家瑞的《伦敦设计周》、冯佳豪的《“寻风来”品牌形象设计》、彭琳的《明珠酒店品牌设计》等作品都展现了设计师的实力。

第十一届IAI获奖作品呈现了IAI前所未有的多元化，参赛作品无疑也提升到了一个更高的水平，但也因此给评审带来更大的挑战。期待未来有更多的优秀设计师参与IAI全球设计奖，分享更多精彩的好作品。同时，更期待IAI竞赛能够为优秀设计师提供更有价值的平台。

2019年06月08日

PREFACE

Compared with the previous sessions, the 2018 IAI Design Award Yearbook has the following new features: the first one is rich in categories, the second one is high in internationalization, the third one is high in gold content, the fourth one is cutting-edge works emerging and the fifth one is the quantity and quality of industrial design products, especially intelligent manufacturing products, have improved significantly. In addition to mainland of China, including Hong Kong and Taiwan, the competition includes 1,376 outstanding works from the United States, Germany, Italy, Japan, Singapore, Switzerland, Greece, Chile, Israel, Malaysia, Ukraine and other countries and regions, 195 works were selected and competed for IAI design awards. After three rounds of judging, 42 awards won the grand prize, including interior, architectural and industrial product design three major categories of awards and outstanding design award,annual best design agency award,special jury award,best environmental friendly award,best material innovation award,best human care award,design star award.Even more gratifying is the emergence of a large number of new faces in the IAI Awards.

In the interior space category, the Japanese designer Hibino's three children's educational space not only reflected design aesthetics in architecture and interior spaces, but also incorporated the concept of humanistic care and humanity in its full of childlike and puzzle space; The"Race Robotics Laboratory"designed by Singapore designer MOD won the IAI Best Design Award - Exhibition Space; Guangzhou designer Huang Yongcai's"Cave Club" used the concept of cave to show that the relationship between people and space, space and space was independent and infiltrated with each other; from Suzhou , a elegant andAnd the"Water Cloud Tea House" from Xu Pin Space who came chic style of tea culture space was presented in a modern Chinese style; Zhang Jianwu & Cai Tianbao (China)'s commercial space work "Layer" was based on its ingenious use of limited spatial structure which cleverly laid out and formed a strong visual impact through the indoor installation and the classic and full of fashion color language of"black and yellow";Jiang Xiaolin (China)'s"Yangshuhua Dream Hotel" began with building, from the landscape to the interior, the soft decoration made a elaborate regulation and design and integrated poetic dream-like natural landscape into space, the works revealed a kind of simple but elegant oriental aesthetic fresh and unique temperament, In addtion, Nayun Interior Design Co., Ltd. (China)'s "Zero Pet Club", American designer Denny Ho's"One Day", Ouyang Kunlun Wei Bing (China)'s "The Space of Wencheng Hall", Creative Group (china Taiwan) " My Playground", Hong Yi'an (China Taiwan)'s "My Secret Garden" and Qian Yinling (China)'s "SEAFOOD CONGEE" all show different spatial aesthetics and traits.

In the architectural design category, the work of chief designer of the Expo Israel Pavilion, Haim Dotan"Zhangjiajie Glass Flyover" adopted an invisible design, which made the flyover disappear naturally in the white clouds, from a distance, the glass bridge was as thin as a cicada's wings, , and it achieved harmony with the natural environment. The bridge created the longest glass facade foot bridge and the steepest strop in the world, the world's first large open space cable suspension bridge and other ten best in the world; Architect Wei Na's work called"Springstream House", the work was inspiring and warm by transforming an abandoned old house and sheepfold on the edge of a small creek in a village in Fujian's eastern Fujian Province, transforming them into a warm-hearted"home" homestay. Kris Lin's architectural design work "Light Waterfall" showcased a distinct personal design style with its consistent sculptural and line-like, bold and innovative space building concept, Swiss Architect Dirk U. Moench's "Huaxiang Church Community Center" showed a community chapel surrounded by shopping centers and office buildings and it had been transformed into a local landmark building. "SLT Office" by AE Architect from Chile was impressive with its comfortable and warm tones, affinity and environmentally friendly materials and green natural elements. Li Shuo's homestay building"King Charm"reflected the local traditional culture and harmoniously integrates the natural ecology with the architectural community, which was amazing for the infinite charm of traditional culture.

In the architectural design category, the work of chief designer of the Expo Israel Pavilion, Haim Dotan"Zhangjiajie Glass Flyover" adopted an invisible design, which made the flyover disappear naturally in the white clouds, from a distance, the glass bridge was as thin as a cicada's wings, and it achieved harmony with the natural environment. The bridge created the longest glass facade foot bridge and the steepest strop in the world, the world's first large open space cable suspension bridge and other ten best in the world; Architect Wei Na's work called"Springstream House", the work was inspiring and warm by transforming an abandoned old house and sheepfold on the edge of a small creek in a village in Fujian's eastern Fujian Province, transforming them into a warm-hearted"home" homestay. Kris Lin's architectural design work "Light Waterfall" showcased a distinct personal design style with its consistent sculptural and line-like, bold and innovative space building concept, Swiss Architect Dirk U. Moench's "Huaxiang Church Community Center" showed a community chapel surrounded by shopping centers and office buildings and it had been transformed into a local landmark building. "SLT Office" by AE Architect from Chile was impressive with its comfortable and warm tones, affinity and environmentally friendly materials and green natural elements. Li Shuo's homestay building "King Charm"reflected the local traditional culture and harmoniously integrates the natural ecology with the architectural community, which was amazing for the infinite charm of traditional culture.

The 11th IAI-winning work presented an unprecedented diversity of IAI, and the entries had undoubtedly increased to a higher level, but it had also brought greater challenges to the jury. I am looking forward to more outstanding designers participating in the IAI Design Award in the future, sharing more exciting and good works, and at the same time, more looking forward to providing a more valuable platform for outstanding designers through the IAI competition.

Oskar

2019.06.08

埃娃·伊日奇娜（捷克）
Eva Jiricna

2018 IAI全球设计奖评委会主席，英国皇家学院院士，美国建筑师学会荣誉院士，捷克文化部终身成就奖。

President of IAI Design Awards Jury, Fellow of the Royal College of England, Fellow of the American Institute of Architects, Czech Ministry of Culture Lifetime Achievement Award.

我对IAI全球设计奖作品的高水准感到惊讶，这些来自不同国家和城市的作品呈现了不同的对当下设计的思考。我很兴奋地看到有很多作品除了重视创意和创新，还非常注重对环保、节能以及人性化的思考。我真的很享受评审的过程，虽然评审让我觉得很困难，因为在如此多的佳作中评出高低是一项挑战。但我认为，无论最终结果怎样，它们能够被分享就已经证明是成功的，我真心地恭喜你们。期待你们未来能够创造出更多好作品。

I am amazed at the high level of the IAI Design Awards. All the works from different countries and cities present different reflections on current design. I am excited to see that many works focus on creativity and innovation, but also on environmental protection and energy saving and human way of thinking. I really enjoyed the review process, although it is very difficult to make a decision with so many good projects of such a high quality., It is a challenge to judge the final high-low level in so many good works. I think, no matter what the final result has proven to be successful in terms of so many good designer have been awarded, I sincerely congratulate you.and look forward to your future creation of many more good works.

朱锫（中国大陆）
Zhu Pei

IAI设计奖学术委员会委员，朱锫建筑设计事务所创建人，中央美术学院建筑学院院长、教授，美国哈佛大学，哥伦比亚大学客座教授APDF国际理事会理事。

Member of IAI Design Award Academic Committer, Zhu Pei Architectural Design Firm Founder, Professor、Dean of the School of Architecture School Central Academy of Fine Arts, Harvard University, Columbia University Visiting Professor Member of the APDF International Council.

IAI全球设计奖的作品具有鲜明的国际化背景。从这些入围的作品中，可以看到全球一体化背景下的设计思维的界限已经越来越模糊，通过不同的表达方式，展现设计者对不同的生活状态和文化背景的思考和理解。IAI全球设计奖的作品内容多元丰富，且有不少可堪称经典之作，值得推广和参与。当然，建筑类别的作品还是偏少。我希望未来能够看到代表这个时代的建筑佳作更多地出现在IAI的竞赛中。与此同时，IAI全球设计奖还在一定程度上增进了不同文化之间的交流，促进了全球设计的理解与合作，并通过竞赛向设计师传递了一种正确的导向，期待下一届IAI全球设计奖有更多佳作。

The works of IAI Design Award have a distinctive international background. From these shortlisted works, we can see that the boundaries of design thinking in the context of global integration have become increasingly blurred and the designers show their thinking and understanding of different living conditions and cultural backgrounds only through different expression ways. The works of IAI Design Award are rich in content and there are many can be called classic works that are worth promoting and participating. Of course, there are still few works of the architectural category. I hope that in the future, I will see more architectural masterpieces representing this era appear in the IAI competition. At the same time, the IAI Design Award has also enhanced the exchanges between different cultures to a certain extent, promoted the understanding and cooperation of global design, and conveyed a correct orientation to the designers through the competition, looking forward to more excellent works in the next IAI Design Award.

陈光雄（中国台湾）
Adam Chen

IAI设计奖学术委员会委员，亚太设计师，联盟（APDF）副理事长，台湾圆镜联合设计事务所设计总监，台湾绿能集团董事长。

Member of IAI Design Award Academic Committee, APDF-Vice Director of Asia Pacific Designers Federation, Design Director of Taiwan Round Mirror Joint Design Office, Council President of Green Energy Group.

IAI全球设计奖的作品每年都有很多变化和特点。在我看来，这也体现了IAI自身的不断创新和升华。今年的IAI作品最显著的变化主要体现在：一是参赛作品的类别比往届多，二是越来越多的设计作品都不同程度地体现了IAI一直倡导的绿色设计理念。比如在室内与建筑的获奖作品中就可以看到设计师已经自觉将绿色环保作为他们作品中不可或缺的元素，包括节能材料、节能设备、节能器具的应用；在工业产品设计的作品中则可以看到可拆卸、可回收、可维护、可重复利用，体现了产品的安全性、节能性和生态性。除此之外，IAI全球设计奖还出现了一个新特点，就是女性设计师的获奖作品的数量快速增加，这一点可以进一步证明未来女性设计师将会挑战一直以来是以男性设计师为主导的天下。

The IAI Design Award features many variations and characteristics every year. In my opinion, this also reflects IAI's own constant innovation and sublimation. The most significant changes in this year's IAI works are mainly reflected in: Firstly, the category of entries has been increased, second ly more and more design works reflect the green design concept that IAI has always advocated in various degree. For example, for the award-winning works of interi or and architecture, designers can con sciously regard sciously regard environmental protection as an indispensable element of their work, including energy-saving materials, energy-saving energy-saving materials, energy-saving equipment, and energy-saving appliances and for the works of industrial products design, four elements including detachable, recyclable, maintainable, and reusable can be seen,reflecting the safety, energy saving and ecology of the products. In addition, a new feature emerging into the IAI Design Award is the rapid increase in the number of award-winning works by female designers, which further proves that women designers in the future will challenge the world that has always been dominated by male designers.

塔莉娅 （意大利）
Dalia Gallico

IAI全球设计奖评委会会员，罗马圣拉斐尔大学时装与设计学院主席，APDF亚太设计师联盟理事成员，ICAA北京国际创意艺术联盟成员。

IAI Design Award Jury Member, Chairman of the School of Fashion and Design, Saint Raphael University.Rome, IAI Global Design Award Jury members, Council Member of the APDF Asia Pacific Designers Federeation, Member of the ICAA Beijing International Creative Arts Alliance.

我很荣幸能够再次作为IAI全球设计奖的评委参加作品评审。IAI全球设计奖是一个誉满全球的设计奖，包括在欧洲和亚洲国家，很多设计师都视参加IAI为荣耀，尤其在欧洲，特别是在意大利，它每年都吸引着世界各国的设计师参加它的竞赛，因为IAI是一个非常具有创新力的奖项，它总是不断地在提升自己和改变自己，从不满足于现状。

I am honored to take part in the review, being a judge of the IAI Design Award again. The IAI Design Award is a world-renowned design award, including in European and Asian countries., many designers regard IAI as a glory. Particularly in Europe, especially in Italy, it attracts designers from all over the world to participate in competition every year. Because IAI is a very innovative award, it is constantly improving itself and changing itself, never satisfied with the status quo.

龙百渡（巴西）
Fernando Brandao

IAI学术委员会委员，龙百渡建筑设计事务所创建人、设计总监，2010年上海世博会巴西馆总设计师，巴西中国友好协会副会长。

Member of IAI Academic Committee, Fernando brandoo architecture & design founder、design director Chief designer of the Brazilian Pavilion at the 2010 Shanghai World Expo, Vice President of the Brazilian China Friendship Association.

我很高兴首次作为IAI全球设计奖评委会的主席参加了作品的评审，我对IAI奖项的认可是因为我看到了它的获奖作品，它们是我迄今为止所看到的最好的设计奖竞赛作品之一。当然，这也因为我曾有幸作为IAI的参赛者并且获得了奖项，所以，我也许比其他评委更了解它，也就更有发言权。IAI全球设计奖最吸引人的一点不是奖项名称，也不是它的那些深受好评的获奖作品，而是它的坚持不懈和敢于自我革新的精神理念，从IAI设计奖的奖杯的创新设计就可以充分证实这一点。值得关注的是，IAI不仅具有高度的全球意识和独特的形象，还有鲜明的评审方向，这对不同国家的设计师来说是相当具有吸引力的。当然，如果还有不足之处的话，就是标志性的建筑作品还不够多。我希望未来能够看到更多来自世界各地的优秀建筑作品出现在IAI设计奖的颁奖典礼和年鉴里。

I am amazed at the high level of the IAI Design Awards. All the works from different countries and cities present different reflections on current design. I am excited to see that many works focus on creativity and innovation, but also on environmental protection and energy saving and human way of thinking. I really enjoyed the review process, although it is very difficult to make a decision with so many good projects of such a high quality., It is a challenge to judge the final high-low level in so many good works. I think, no matter what the final result has proven to be successful in terms of so many good designer have been awarded, I sincerely congratulate you.and look forward to your future creation of many more good works.

白福瑞（瑞士）
Florin Baeriswyl

IAI 学术委员会委员 ,APDF 国际理事会理事，DAI（Design, Architecture,& Identity），品牌顾问公司设计总监、创办人。

Member of the IAI Academic Committee, APDF International Council Director,DAI (Design, Architecture, & Identity),Brand Consultant Design Director、Founder.

IAI全球设计奖是一个正处在上升期的国际品牌，它经历了12年的沉淀和塑造，如今已经成长为翩翩少年，现在即将迈向更富有活力、更成熟的青年阶段。这是一个属于中国的本土国际品牌，我参加了很多次IAI评审的活动，我看到了它每年的逐步成长，其影响力不断扩大，作品已经覆盖了所有设计领域，这是我在中国所接触到的第一个如此完整的专业设计奖项。而且它已经进一步完善了品牌的定位，设立了IAI室内奖、IAI建筑奖、IAI智造奖，在传承原有品牌视觉形象的同时又做了创新，成功实现了品牌的转型和升级。这种创新代表了与时俱进，也代表了IAI主办方的态度，这是我愿意一直参与它的理由。

The IAI Design Award is an international brand that is on the rise. After 12 years of precipitation and shaping, it has grown into an elegant young man and is now moving towards a more dynamic and mature youth.This is a Chinese local international brand. I have participated in many IAI review activities, I have seen its annual growth and influence enlarging, the works have already covered all the design fields and this is the first so complete professional design award that I have encountered in China. Moreover, it has further perfected the positioning of the brand, setting up the IAI Interior Award, IAI Architecture Award, IAI Intelligent Manufacturing Award and it innovates while inheriting the original brand' s visual image, successfully realized the transformation and upgrade of the brand. This kind of innovation represents the advancing with the times and the attitude of the IAI organizer. This is the reason why I am willing to participate in it all the time.

李淳寅（韩国）
Soon-In Lee

APDF亚洲区理事长（2018-2020），首尔设计中心负责人，IAI学术委员会会员。

Asia Pacific Designers Federation (APDF) Council director of the Asian Region (2018-2020), Head of Seoul Design Center, IAI Academic Committee Member

每次参加IAI全球设计奖作品的评审总是让我感到紧张，因为我常常会为那些精彩的作品而纠结，担心投票的结果有时候不能如我所愿。当然，评审的最终结果似乎总是众望所归，如同你此刻看到的结果一样。IAI设计奖在未来需要更多来自产品领域的作品参赛，尤其在智能智造方面，我认为IAI全球设计奖已经完成了IAI设计奖品牌的建立，现在开始进入品牌进化的阶段。值得欣慰的是，IAI已经与时代同步，IAI智造奖的设立就是一个有说服力的证明。期待下一届IAI有更多好设计！

Every time I participate in review of the works of IAI Design Award, I feel nervous, because I often get entangled in those wonderful works, worrying that the results of voting sometimes can't follow my wishes. Of course, the final result of the final review always seems to be in favor with the general public, just like the results you see at the moment. IAI's award-winning works will require more design works from the product field in the future, especially in terms of intelligent manufacturing. I think the IAI Design Award has already completed the establishment of the IAI Design Award brand and now it is entering the stage of brand evolution. What is gratifying is that IAI has already kept up with the times, and the establishment of the IAI Intelligent Manufacturing Award is a convincing proof. Look forward to the next IAI emerging more good design works!

杰斯·斯班杰斯（荷兰）
Kees Spanjers

APDF联合主席，前欧洲室内设计师联盟（ECIA）主席，IFI 理事会成员（2004-2009）。

Co-Chairman of the Asia Pacific Designers Federation(APDF), Fomer President of the European Interior Designers Association (ECIA),Member of the IFI Council (2004-2009).

我自从2012年第一次参加IAI设计奖的评审，迄今已经参加了三次。每次参加IAI评审都可以看到IAI参赛作品在质和量上发生的变化。显而易见，IAI的作品越来越国际化，参赛者水准越来越高，对我的评审要求也越来越多，因为我需要凭借我的经验、专业知识和借助我的肉眼来给每个作品打分，我感觉越来越难做出最后的决定，尤其在最终锁定的那些大奖的候选名单里。IAI设计奖就像一面镜子，反映了当今世界设计界的现状，正如中国的国力一样蒸蒸日上，IAI设计奖也正在向更高的目标迈进。我衷心祝愿IAI能够再接再厉，立足中国，借助中国的传统文化、东方的美学思想，汲取世界文明的精华，让IAI成为最具含金量和品牌价值的世界性设计大奖。

I have participated in review of the IAI Design Award for the first time since 2012 and have participated three times so far. Every time I participate in the IAI review, I can see the changes in the quality and quantity of the IAI entries. Obviously, IAI's works are more and more international, the level of the participant is getting higher and higher and the adjustment to me is naturally getting bigger and bigger. Because I need to rely on my experience, professional knowledge and my naked eyes to score each piece, I feel more and more difficult to make the final decision, especially in the final locked list of candidates of the grand prize. The IAI Design Award is like a mirror reflecting the current situation of the design community world, just as China's national strength is flourishing and moving toward higher goals. I sincerely hope that IAI will continue to make persistent efforts, be established in China and rely on Chinese traditional culture, oriental Aesthetic thinking, draws on the essence of world civilization and makes IAI become the design award worldwide with the highest gold content and brand value.

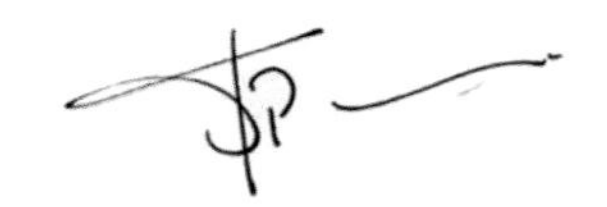

Commercial Space

室内 —— 商业空间

一 天
ONE DAY

项目地点：中国成都 / Location : Chengdu, China
项目面积：112平方米 /Area :112 m²
设 计 师：何丹尼 / Designer : Denny Ho

IAI 最佳设计大奖 I IAI 最佳人文关怀大奖

何丹尼 Denny Ho

美籍华人室内设计师，毕业于加州圣安娜艺术学院，壹阁设计工作室创始人。
东京艺术大学进修生、创基金 B 计划荣誉学员、米兰理工大学艺术设计双硕士学位、中国香港室内设计师协会荣誉会员、40 UNDER 中国（四川）设计杰出青年、成都装饰协会会员、中国装饰协会同盟会会员、CCTV-2 交换空间常驻设计师、多乐士官方签约设计师。14 年专注工业、室内、家居设计。

American Chinese interior designer,Graduated from the Art Institute of California-Santa Ana, founder of ONE SPACE DESIGN.
Tokyo University of the Arts, advanced student,Honorary student of the Foundation B Project,A double master's degree in art design from the Politecnico di Milano and an honorary member of the Hong Kong Interior Designers Association.
40 UNDER China (Sichuan) Design Outstanding Youth, Member of Chengdu Decoration Association,Member of the China Decoration Association League, CCTV-2 exchange space resident designer, Dulux official contract designer. 14 years of focus on industrial, indoor, home design.

空间是会说话的，这并不是伪命题。设计师何丹尼在这套作品中，实现了他对共享空间的完美畅想。

《一天》是位于成都的一家时尚买手店。整个空间里充满未来主义的设计风格，令人错以为置身于不一样的空间世界。

在这里，艺术元素从各个角度融入，使逛店的过程充满享受。这些富有故事的艺术品与华美的服饰相互映衬交织，就像一条隐形的脉络，等着你一个个去找寻，充满诱惑。

店主雨晴是一名90后，她也是常年往返于欧洲和中国的时尚达人，多年在国外的经历让她的审美观念越发大胆和包容。她希望自己的买手店独一无二，有趣、有活力，并且能让客人一走进她的买手店就留下深刻的印象。为雨晴创造这样的一个有趣的、具有视觉冲击力的空间是让设计师在设计过程中也很享受。温暖高饱和的爱马仕橘色让整家店充满热情和活力，年轻又不失质感。整个空间展现了90后的时尚和生活态度，充分展现了她的个性、想象力与表现力。

Space is talking, this is not a false proposition. Designer Denny Ho realized his perfect imagination of the shared space in this work.

One Day is a fashion buyer's shop in Chengdu. The whole space is full of futuristic design style, which makes people feel that they are in a different space world.

Here, the integration of artistic elements from all angles makes the process of shopping full of enjoyment. These rich-stories artworks and colorful costumes are intertwined with each other, like an invisible vein, waiting for you to find one by one, full of temptation.

Yuqing, after ninty. She is also a fashionista who travels to Europe and China all the year round. Her years of experience abroad have made her aesthetic concept bolder and more inclusive. She wants her own store to be unique, fun, energetic, and to impress guests as soon as they walk into her store. Creating such an interesting and visually striking space for Yuqing is also very enjoyable in the design process. Warm and highly sqturated Hermes orange makes the whole store full of enthusiasm and vitality, young and without losing texture. The whole space shows the fashion and life attitude after the 90s, fully demonstrating her personality, imagination and expressiveness.

一天从日出温暖的橙色开始，轻快愉悦。从黎明慢慢走到黑夜，享受完美一天。运用色彩的渐变，营造从清晨到日暮的光影变化。

大胆的现代呈现，博物馆式的陈列方式，材质与灯光的运用极富高级感，让空间与人互相达到平衡，给展示品最大的发挥余地。

提升整个空间的艺术感和设计氛围，显现一贯以来的细腻质感的同时，把色彩对空间品质的坚持体现得淋漓尽致，预示着这里将生发出更多新鲜时尚与可能性。

将人引入一处极具迷幻感的空间，色彩饱和、材质混搭，亮面及金属线条镶嵌其中，混合得极为完美。既有张力，又不失时尚。

One day begins with the warm orange of the sunrise, brisk and pleasant. From the dawn to the night, enjoy the perfect day. Use the gradient of color to create light and shadow changes from morning to nightfall.

The bold modern presentation, the museum-style display method, the use of materials and lighting are highly sophisticated, allowing space and people to balance each other, without any sense of contradiction, giving the showpiece the greatest scope.

Enhance the artistic sense and design atmosphere of the whole space, and present the consistent exquisite texture. At the same time, the persistence of color towards space quality is reflected incisively and vividly, which indicates that there will be developing more fresh fashion and possibilities.

Introducing people into a space with whimsical sense , saturated colors, blended materials, shiny surfaces and metal lines are embeded in and perfect for mixing. With tension and without losing the fashion sense, the unique style of the scene is fascinating and gives afford for thought.

层 迭
Layer Up

项目地点：中国福建 / Location : Fujian, China
项目面积：110平方米 / Area :110 m²
设 计 师：张健武 & 蔡天保 / Designer : Zhang Jianwu & Cai Tianbao

张健武 & 蔡天保
Zhang Jianwu & Cai Tianbao

2019 德国 IF 设计奖
2018 意大利 A' 设计大赛银奖
2018 德国红点奖的红点设计大奖
2018 第十一届台湾室内设计大奖的商业空间 / 休闲空间 TID 奖
2017 中国设计品牌大会最具投资价值商业品牌空间
2017 第八届中国国际空间设计大赛商业、金融、售楼处空间 / 工程年度创新作品奖

2019 German IF Design Award
2018 Italy A'Design Award Silver Award
2018 German Red Dot Award Red DOT Design Award
2018 The 11th Taiwan Interior Design Award Commercial Space/Leisure Space TID
2017 Award
China Design Brand Conference, the most investment-worthy commercial brand space
2017 The 8th China International Space Design Competition Commercial, Finance,
Sales Office Space/Engineering Annual Innovation Works Award

由于空间结构限制，一到二层的层高过高又不能增加夹层，所以首先要解决两个问题：第一是保证功能需求的前提下，不让空间显得狭长；第二则是楼梯的位置。将两个问题重叠，会发现楼梯的位置和动向即为本案设计的重点。

解决了结构布局方面的问题之后，如何把装置烘托成视觉上的焦点，成了设计师另一个需要思考的问题。“黑+黄”是非常经典的颜色搭配，既有时尚意味又有强烈的视觉冲击力。没有多余的造型堆砌，用最基础的结构，通过颜色的对比，把楼梯的装置感体现出来，使之成为空间视觉焦点。

Because of the limitation of spatial structure, the first layer is too high and the interlayer can not be added, so it is necessary to solve two problems: the first is to keep the space narrow and narrow while the second is the location of staircase. Overlap the two questions and find that the location and movement of the staircase is the focus of the case.

After solving the structural layout problem, how to turn the device into a visual focus has become another problem that designers need to think about. Black and yellow is a very classic color match, both fashion and strong visual impact. Do not pile up extra shapes, with the most basic knot Structure, through the contrast of colors, the installation of the staircase sense of reflection, become the focus of space vision.

水云间茶会所
Water Cloud Tea House

项目地点：新疆 / Location : Xinjiang , China
项目面积：460平方米 / Area : 460m²
设 计 师：蒋国兴 / Designer : Jiang Guoxing

IAI 杰出设计大奖

一茶一世界，一味一人生。人生如茶，第一道茶苦若生命，第二道茶香似爱情，第三道茶淡如清风。设计师从周敦颐读书修行的月岩洞、曾国藩故居及徐志摩的《再别康桥》汲取创作灵感，以现代中式设计手法打造出禅意清幽、素雅别致的茶文化空间。

One tea, one world, one life.Designer drawed inspiration from Zeng Guofan's former residence and Xu Zhimo's "On leaving Cambridge" .Designer used modern Chinese style to build a beautiful zen, simple but elegant China's tea culture space.

蒋国兴 Jiang Guoxing

叙品空间设计有限公司董事长兼设计总监，他1996 年毕业于厦门工艺美院。从业二十余年，他始终坚持原创设计，并将禅文化运用到空间设计，从而使空间更有深度。
凭借对原创的坚持和对东方文化的理解，获得了很多国内外的设计荣誉。如英国安德鲁马丁、意大利 A' 设计大奖、德国 IF 设计奖等。

Xupin Space Design Limited Company's Chairman&Design Director,He graduated from Xiamen institute of arts and crafts in 1996 and has been working as a designer for more than 20 years.He always stick to original design ,meanwhile he apply Zen to space design which make the space more thoughtful and appealing.With the persistence of original design, the understanding of Oriental culture,Mr Jiang won a lot of honners in both home and abroad. Such as Andrew Martin Interior Design,Italian A' Design Award ,Gremany IF Design Award and so on.

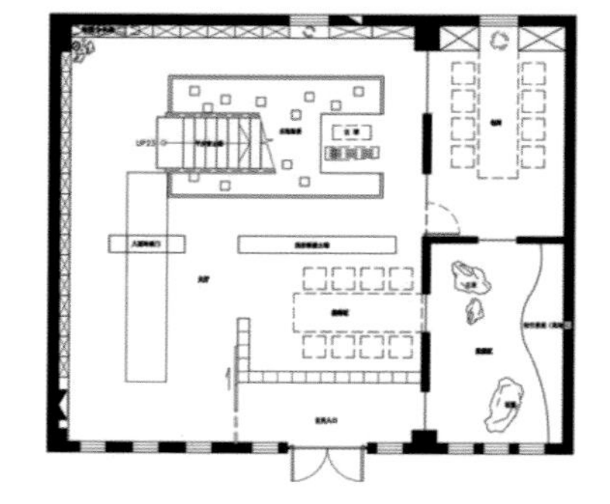

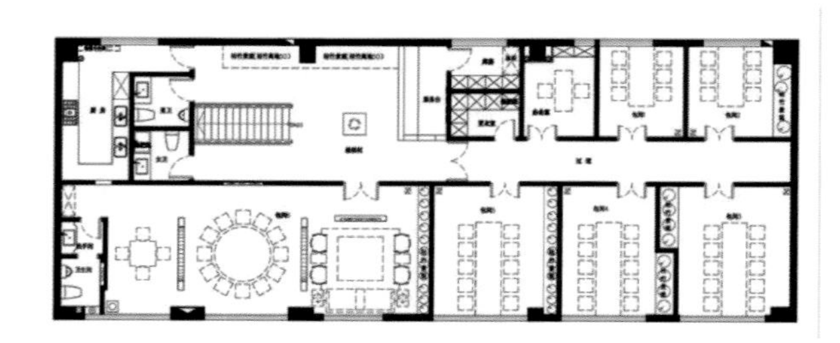

悄悄的我走了，正如我悄悄的来；
我挥一挥衣袖，不带走一片云彩。
Very quietly I left，
As quietly as I came here;
Gently I flick my sleeves，
Not even a wisp of cloud will I bring away.
—— 徐志摩
Xu Zhimo

水雲間

水雲間

年年春自東南來
建溪先暖水微開

茶所
Tea Here

项目地点：中国北京 / Location : Beijing, China
项目面积：140 平方米 / Area : 140 m²
设 计 师：欧阳昆仑、魏冰 / Designer : Ouyang Kunlun , Weibing

欧阳昆仑 Ouyang Kunlun

1994 毕业于重庆建筑工程学院
2002-2004 人民大学徐悲鸿美术学院设计艺术系空间设计专业客座教师
2003- 至今 任北京方和建筑及室内设计总监

Graduated from Chongqing Institute of Civil Engineering and Architecture in 1994.
2002-2004 Department of Design Art, Xu Beihong Academy of Fine Arts, Renmin University Space Design Professional · Guest Teacher.
2003-present Beijing Fanghe Building and Interior Design Director.

魏冰 Wei Bing

1995 毕业于中央工艺美术学院环艺系环境设计专业学士学位
2003- 至今 任北京方和室内设计总监

Graduated from the Central Academy of Arts and Crafts in 1995 with a bachelor's degree in environmental design.
2003-present Beijing Fanghe Interior Design Director.

穿过陶瓷街上的一片干竹，开门见山是“亦吾安”园。砖和土的山水基调延伸至“茶所”内部。

这是一片开放、舒适的当代公共天地。物尽其用、空间无尽其用是其设计宗旨。

Keep going on Ceramic One street, behind a bush of dried bamboo is the gate to Tea Here. Pushing open the door, first thing you see is a miniature garden – Yi Wu An Garden, a primarily brick and soil structure that extends into the interior of Tea Here. This is to mimic the ancient Chinese way of living: open the door, and you see mountains.
Inside Tea Here is an open, comfortable and contemporary space, designed with the motto of ‘making the best possible use of materials yet leaving the space open to be used whatever way you want’ .

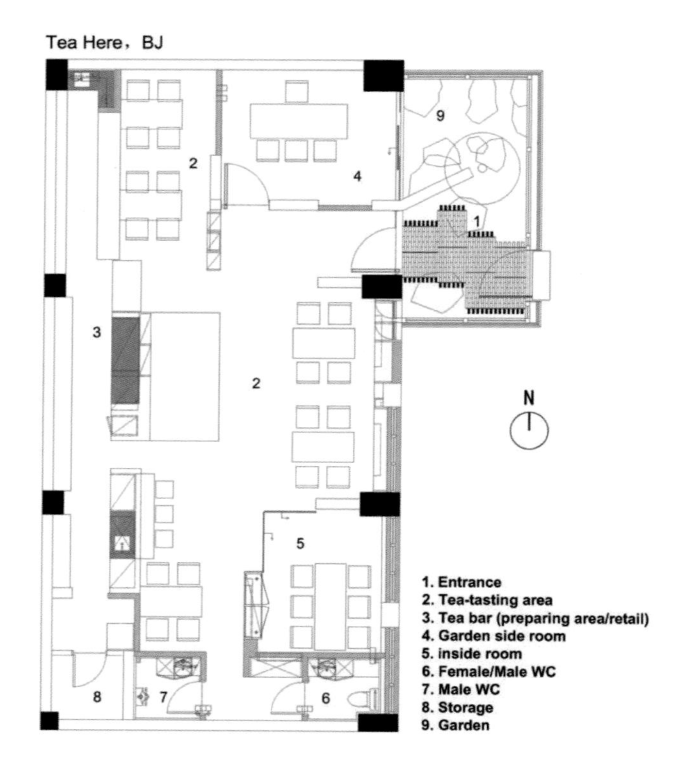

利用20世纪50年代厂房结构，将各时代遗留的灯架、门扇等材料，建造痕迹保留；并将旧物与今天的空间结合（原空间里砖墙拆除后亦用来砌筑新茶所的 VIP 房）。

混构让我们得到新生即老旧的、充满时间气息、又与今天保持了距离的茶馆。

Making The Best Use Of Materials

We built Tea Here within the remains of a 1950s factory. We preserved the original lamp holder, door leaf, other materials and old construction mark, and mixed them with contemporary pieces. For instance, the original bricks were demolished and re-used as the partition of the VIP rooms.

The mixed use of old and new allow us to feel the new came from the old, that there is a lineage of time and within the Tea Here space, we can still keep a distance from today.

空间无尽其用。
茶所不大：艺术介入的男女用卫生间、储藏室、两间 VIP 室。
及六组（即文初所说的“亦吾安”园）。
竹与园让我们得到与商街保持了距离的茶馆。

Creating A Limitless Space

Tea Here is not big, but art is manifested in every corner including the uni-sex bathroom, storage room, two VIP room, the six sitting areas surrounding the huge tea counter that runs through from the north to the south end of the space, as well as the 200 tea placard hanging over the counter.

Albeit small, Tea Here creates a garden at the entry way, i.e., Yi Wu An Garden. You can look at it, walk around it and even sit in it. Bamboo and Garden create a distance between the Tea Here space and the commercial street outside.

茶

入“茶所”亦是入园，入一方自足天地。茶所的存在补充完善了“798”的艺术生茶所的存态，它是茶圈的艺术空间，亦是艺术圈的茶空间。

Entering Tea Here is entering a garden, entering a space with seclusion and self sufficiency.
The creation of Tea Here complements the artistic eco-system in 798 Art District. It is an artistic space for people who enjoy tea, as well as a tea house for the artists.

入"茶所"亦是入园，入一方自足天地。

ENTERING TEA HERE IS ENTERING A GARDEN,
ENTERING A SPACE WITH SECLUSION AND SELF SUFFICIENCY.

零宠物俱乐部
Zero Pet Club

项目地点 : 中国南京 / Location : Nanjing, China
项目面积 : 475 平方米 / Area : 475 m^2
公司名称 : 拿云室内设计有限公司 / Organization Name : Nayun Design

IAI 最佳设计大奖

南京拿云室内设计有限公司成立于 2013 年。公司成立伊始，即致力于商业空间设计的探索与实践，核心的设计团队有着不同的教育和实践经验，使作品呈现出更加丰富多元的创意。

设计领域包括商业空间设计、办公空间设计、商业地产等。作为一个有着丰富经验的设计团队，拿云设计有着自己的设计风格与态度。用创新的激情在业界赢得了口碑和荣誉。

Nanjing Nayun Interior Design Co., Ltd. was founded in 2013, which is devoted to the exploration and practice of commercial space design. The core design team has different education and practical experience, which makes the works show more rich and diverse creativity. Design areas include commercial space design, office space design, commercial real estate, etc. As a design team with rich experience, Nayun Design has its own design style and attitude. With the passion of innovation, we have won praise and honor in the industry.

零宠物俱乐部位于南京市玄武区天山路39号，开放而明亮的门头设计犹如无声的邀请，吸引着人们将视线自然而然地延伸至店面内部。乍一看，会以为这是一家书店或者咖啡馆，然而外墙上被设计成猫咪图案的灯牌早已明确宣告了这个空间真实属性的。

与常规宠物店相比，零宠物俱乐部其实更像是一个宠物社交生活馆，在这个475平方米的空间里，主要进行一些宠物的行为矫正训练以及经营其他宠物周边产品，比如宠物旅游、保险、写真等。通过对这些服务的贩售，零宠物俱乐部希望能进一步拉近人和宠物之间的距离。

Zero Pet Club lies at the 39th Tianshan Road , Xuanwu District of Nanjing, whose gate is open and shining just like a silent invitation so that the attention of visitors are all attracted to its inner space naturally.At a glance, we may recognize it as it is a book store or coffee shop, but the light boards hung on the outside wall were designed to a cat pattern had proclaimed the real feature of this space.

Compared with common pet club,actually, Zero Pet Club is more like a pet social life club. In this 460-square meter area, the behavior of pets will be corrected and pet related products will be sold such as pet travel, pet insurance , pet portrait and so on. Through selling these series of services, Zero Pet Club hope this can make human and their pets closer and communication further.

最开始，设计师对这个项目的定位就是“巧而美”：以白色为基调，打破一般宠物店狭小拥挤、不易通风的格局，通过极简主义的设计手法，搭配原木、玻璃、钢材、裸管等归本元素，在高楼林立的繁华都市中，打造一个具有复合之美的宠物天地。大厅是一个综合性区域，宽敞磊落的的平面布局、冷暖适宜的灯光效果、浓淡结合的色彩处理，都十分具有流动感和趣味性。宠物休息区用原木和玻璃分隔出一个个独立的“房间”并用数字标记以方便管理，“房间”前的小沙发凳可以灵活移动，主人可以坐在透明的玻璃窗口和宠物互动，同时又不打扰其他房间的“住客”。猫咪玩耍区有着强烈的自然风格，各种空中爬梯、猫爬架组成一个冒险乐园，独立的设计既可以让小家伙们玩得开心，又不至于抢占大厅的其他公共空间。

From the very beginning, designers attempted to design this project " exquisite and beautiful": white as keynote, breaking form of the common pet shop where inner space is crowd and has bad ventilation,through minimalism design techniques, suiting logs,glasses,rolled steel,and bare pipe such kinds of original elements to build a pet world with compound beautiful in this city that full of tall buildings and towers.The hall is a comprehensive area. Wide and open plane layout,appropriate light effect and shades affordable color design that they all possess sense of flow and interest. Pet rest rooms are separated by logs and glasses and signed at numbers in order to manage. Small sofa and bench can be mobile anywhere so that owners can sit at the front of the glasses to communicate with their pets,at the same time, they also wouldn't disturb guests in other rooms.Cat playing area has strong natural style,in which many kinds of air ladders and cat climbs consist a adventure park. Single design not only let cats have fun,but also take up other space of the hall.

观整个空间，设计师对白色和原木色的娴熟使用使得零宠物俱乐部成为一家极具个人风格特色的宠物店，并且很好地表现出品牌的温馨度和精致感。顾客来到这里，感受的不仅仅是专业和放心，更多的是零宠物俱乐部对宠物真切的关怀、喜爱和照顾。

Looking this whole space,designers let Zero Pet Club become a personal style pet shop through take advantage of white and wood to express the worm and exquisite of the band. When customers come here,they feel not only release but professional. What's more, they can get sincere affection and care to pets from Zero Pet Club.

ZERO PET CLUB
COFFEE
ZERO PET CLUB

Birman
Russian Blue
Sphynx

K11购物艺术中心
K11 Shopping Mall Artistic Centre

项目地点 : 中国广州 / Location : Guangzhou, China
项目面积 : 35000 平方米 / Area : 35000 m²
公司名称 / Organization Name : LTHK Group

IAI 设计优胜奖

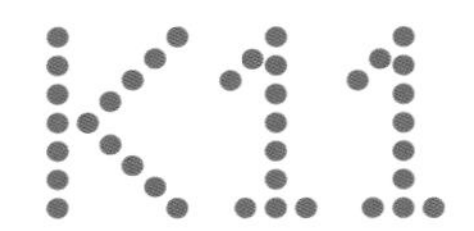

K11

K11 是新世界发展有限公司旗下的高端生活品牌，是首个把艺术、人文、自然三大元素融合为核心的全球性原创品牌。

K11 is a high-end lifestyle brand under the New World Development Co., Ltd., and is the first global original brand that integrates the three elements of art, humanity and nature.

LTHK Group

LTHK Group 成立于 1998 年，在中国香港及广州、上海均设有办事处，是由中国香港注册建筑师谢宏准先生及杨家声先生主导的设计事务所。

Established in 1998, LTHK Group has offices in Hong Kong, Guangzhou and Shanghai. It is a design firm led by Hong Kong registered architects Mr. Xie Hongzhun and Mr. Yang Jiasheng.

位于广州中央商务区内、毗邻五大文化地标，楼高530米的周大福金融中心是目前全广州最高的商业地标，而位于周大福金融中心裙楼的广州K11购物艺术中心是珠江新城中轴线上仅有的高端商业综合体项目，独一无二的零售美术馆概念将引领广州体验商业的新变革。

Located in the Central Business District of Guangzhou, adjacent to the five major cultural landmarks, the Chow Tai Fook Financial Center is 530 meters high and is currently the highest commercial landmark in Guangzhou. The Guangzhou K11 Shopping Art Center in the Podium of Chow Tai Fook Financial Center is the high-end commercial complex projects that only exits on the central axis of Zhujiang New Town. The unique concept of retail art museums will lead the new revolution in experience business in Guangzhou.

金色是广州K11购物艺术中心的主体色调，也是K11品牌的主题颜色。从B1中庭，到各层背景墙，再到天花板，随着空间的变化，柔和的金色仿佛溪流般流淌至空间的每个角落，突出项目的奢华高雅与其独特的艺术氛围，并强化人们对K11品牌的认知。

室内农庄是整个项目最为独特之处，内设大型生态互动体验种植区，突破室内环境的局限，并采用多种高科技种植技术模拟蔬菜的生长环境，让消费者彷佛沉醉于城市的绿洲，唤起他们珍惜大自然馈赠、保护城市生态的意识。

随处可见的装饰树枝与树叶元素象征与南粤文化密不可分的榕树。榕树苍劲挺拔，生命力顽强，自然在南方人民心中生成一种特有的“榕树文化”。项目以榕树为设计主线，不仅把小区、本土艺术、文化元素融于设计之中，让消费者产生亲切感，还将艺术赏析、人文体验与自然环保三者完美结合，缔造独特而充满革新的多元文化商业空间。

Gold is the main color of Guangzhou K11 Shopping Art Center, and it is also the theme color of K11 brand. From the B1 atrium to the various background walls of each floor and ceilings, as the space changes, the soft gold flows like a stream to every corner of the space, highlighting the luxury and elegance of the project and its unique artistic atmosphere, and strengthening people's K11 Brand awareness.

The indoor farm is the most unique part of the whole project. It has a large-scale ecological interactive experience planting area, breaking through the limitations of the indoor environment, and using a variety of high-tech planting techniques to simulate the growing environment of vegetables, so that consumers seem to be immersed in the oasis of the city, evoking the awareness of cherishing the gift from the nature and the protecting of urban ecology.

The decorative trees and leaf elements that are seen everywhere can symbolize the banyan trees that are interwoven with Nanyue culture. The banyan trees are vigorous and forceful, and its vitality is tenacious. Naturally, a unique "banyan trees culture" is created in the hearts of the people of the South. The project takes banyan trees as the main line of design, which not only integrates the community, local art and cultural elements into the design, in which gives consumers a sense of intimacy, but also perfectly combines artistic appreciation, humanistic experience and natural environmental protection this three to create a unique and innovative multivariant cultural business space.

凡舍文创
Fanshe Bookstore

项目地点：中国四川 / Location : Sichuan, China
项目面积：600 平方米 / Area : 600 m²
公司名称：Woood空间设计 / Organization Name : Woood Interior Design

IAI 设计优胜奖

凡舍书店整体设计将以蒙德里安的艺术画为元素进行的构造延用到整体空间上，从室外门头到室内的延伸，由于整体地势条件的优势位于两条道路交汇处，但是实际上在展示上有很大的问题，因此我们用了方形元素阵列的模式，凸显空间本身的视角，让整体空间更为抢眼，每一处的室外建筑窗户设计也是有独到之处，用了相框的元素、适当的切角，让所有室内空间都框于相框之内，处处营造景观。

在功能上，空间前半部分考虑的是展示和阅读区，后半部分考虑的是沙龙和简餐的区域，功能划分明确，但是考虑到整体层高不是很高，我们只是在局部搭建了一些空间，在下方的空间中为了不那么压抑，运用了镜面的效果，从而降低视觉压力，达到一种平衡。整体空间在提升商业用途上，用植物和装置艺术的表达方式让整体空间达到一种生态和谐的空间环境，以达到人与空间、人与自然和平相处的意向。

The overall design of the Fanshe Bookstore we have studied the structure of the elements of Mondrian's art painting and extended it to the whole space. From the outdoor door extending to the interior, the advantages of the overall terrain conditions are located at the intersection of two roads, but actually there are big problems in the display, so we used the square element array mode to highlight the perspective of the space itself, making the overall space more eye-catching. Every outdoor building window design is also unique, using the elements of photo frame , the appropriate cut corners, which allow all the interior spaces to be framed inside the frame, creating a landscape everywhere.

In order to reduce the pressure on the space below, the mirror effect is used to reduce the visual pressure and achieve a balance. The overall space is used to enhance commercial use, and the expression of plants and installation art allows the overall space to achieve an ecologically harmonious space environment. In order to achieve the intention in which people and space, people and nature live in peace.

西塘玺樾销售中心(软装陈设)

Xitang Xiyue Sales（Soft Decoration Display）

项目地点:中国浙江 / Location : Zhejiang, China
项目面积:686 平方米/Area : 686 m²
公司名称：邸设空间设计（上海）有限公司
Organization Name : Decor & Decor Interior Design (Shanghai)

IAI 设计优胜奖

邸设设计 Decor & Decor Interior Design (Shanghai)

成立于2015年，由从业十几年的国际国内软装主创共同打造的国际设计团队，设计师主要是业内顶尖女性设计师为主，主打室内硬装、软装设计与产品改造，品牌秉承着对精致生活的追求，打造室内软装设计结合家居产品私人定制设计为一体的软装设计模式。在为居住者呈现更舒适、更精致的室内设计的同时，融入居住者个人性格与形象气质的私人定制空间。

Décor & Décor Interior Design Ltd.,Co was founded in 2015, the international design team created by the international and domestic soft-wearing designers who have been engaged in the industry for more than ten years, the designers are mainly the top female designers in the industry, focusing on the interior hard-wearing, soft-packing design and product transformation. The brand adheres to the pursuit of exquisite life and builds The soft design of the interior is combined with the custom design of the home product. For the occupants to present a more comfortable, more refined interior design, while incorporating the personal custom of the occupants' personal character and image.

硬装在周围使用了干净的大理石，衬托了一个安静高雅的氛围，顶面是卷轴一角配上木饰面让中堂的茶桌成为了这个空间中最耀眼的明星，四周展开的湖平面画卷让人联想到平静的海上泛起的一叶扁舟，因此在茶桌的造型上，我们选择了酷似帆船的形态，在底部用琉璃作为支撑，让整个造型像悬浮在海平面上的一叶扁舟。四周搭配的琉璃长椅让整个空间更加灵动鲜活。

在此次的设计中，琉璃的元素随处可见，琉璃是采用古代青铜脱蜡铸造法高温脱蜡而成形。经过十多道手工工艺的精修细磨，稍有疏忽即可造成失败或瑕疵。琉璃内，或大或小、或沉或浮的气泡，是琉璃生命的特征。这些气泡游走于晶莹剔透的水晶之中，漂浮于柔情似水的颜色带之间 。是快意洒脱， 是情意绵绵， 还是浩然气魄。琉璃在述说，我们在倾听。愿我来世，得菩提时，身如琉璃，内外明澈。

Hardware decoration uses clean marble in surroundings to set off the quiet and elegant atmosphere. The roof is the scroll matching with the wood facing to make the tea table become the most dazzling star in this space. The lake level scroll spreading in surroundings makes me associate with a tiny boat on the clam ocean, thus the tea table modeling is similar to the sailboat. The bottom is supported by the glass, thus the entire modeling forms the concept of a boat suspending in the seal level. The glass bench in surroundings makes the entire space become flexible and alive.

In this design, glass elements can be seen everywhere. Glass applies ancient bronze lost wax casting method to from crystal works through the high-temperature dewaxing. After dozens of fine trimming processes, any negligence will result in failure or defect. In glass, more or less, sunken or floating bubbles are features of glass life. These bubbles are wandering in clear crystals and floating in the tender and soft color belts. It is free and easy, emotional and spacious. Glass is telling, while we are listening. I hope that we will be clear as glass in the future world.

玺樾西塘

木里木外展厅(软装陈设)

WoodStory Exhibition Hall （Soft Decoration Display）

项目地点：中国江苏 / Location : Jiangsu, China
项目面积：315 平方米 /Area : 315 m²
公司名称：邸设空间设计（上海）有限公司
Organization Name : Decor & Decor Interior Design (Shanghai)

IAI 设计优胜奖

凡人必有七情：

喜、怒、忧、思、悲、恐、惊。

六欲：

眼、耳、鼻、舌、身、意。

空间：

便是承载七情六欲的地方。

于是，

邸设在木里木外里用东方的正五色：

至诚、至简、乐山、乐水、无为，

开启你奢华绚丽的五彩生活。

All mortals have seven emotions:
pleasure, anger, worry, anxiety, sorrow, fear and astonishment;
and six sensory pleasures:
eyes, ears, nose, tongue, body, and mind;
while space is the container to carry them.
Thus,
we set this space in the iWoodStory Exhibition Hall. With adoption of the five positive colors in ancient oriental cultural concepts: green, red, white, yellow and black, our product series of extreme sincerity, extreme simplicity, mountain fancy, water fancy, quietism will bring you a sumptuous, colorful and gorgeous life!

茶阁里的猫眼石

Opals In Teahouse

项目地点：中国四川 / Location：Sichuan，China
项目面积：80 平方米 / Area：80 m²
公司名称：创空间 / Organization Name：Creative Group

IAI 设计优胜奖

創空間
CREATIVE GROUP

汇集建筑与室内设计背景的多元人才，配合组织化管理模式，将工程进度做系统化的全面控管，提供从前期沟通，设计，合约拟定到后期执行的完整服务。2000 年成立“权释设计”，2014 年，2015 年陆续成立“CONCEPT 北欧建筑”、与“JA 建筑旅人”建筑设计公司，传递人与自然和谐共生的理念精神，秉持着“提升国人生活品质、共创生活美学体验”为目标持续优化。

Bringing together diverse talents in the background of architecture and interior design, and coordinating the organizational management model, systematically and comprehensively control the progress of the project, providing complete services from pre-communication, design, contract formulation to post-execution. In 2000, the company established the "ALLNESS Design" . In 2015, it established the "CONCEPT " and the "Journey Architecture" architectural design company to convey the spirit of harmony between Humanity and Nature Environment, and uphold the "quality of life of the people." is the continuously optimized goal .

以顾客体验出发，从茶香开始，感受茶阁空间，创造符合在地特色的店铺，带给顾客令人感动的体验。设计风格简约、怀旧、工业风。在地化仿古设计，并与张爱玲母校（圣玛莉亚女校）做内外呼应的空间情境设计，外观造型融合在室内空间中，以相同质感、怀旧的色调灯光，仿佛将大家带回20世纪30 年代的老上海，喝一口茶、饮一口历史的痕迹。

The case features one store ,one design concept, and the team shapes the different styles of the teahouse in response to the style of local store’ s surroundings to express the brand story that can be a place for consumers to linger again and again . The antique design Santa Maria's Girls' School is designed to coordinate the interior and exterior, and the exterior is integrated into the interior space.With the same texture, nostalgia and old-fashioned shades of light, it seems as if you will bring everyone back to the old Shanghai in the 1930s, drink a cup of tea and drink a trace of history.

画框式观景窗，特别取材自女校外的哥德式窗框，让消费者由内而外，看见室内就是一幅画。分族群的饮茶空间 ，能巧妙使用破格空间，将一人座、情人座、三五好友座位分隔，同时满足各族群的聚会需求，让人能够自在地在空间找到最适合的饮茶角落。星空垂坠吊灯点缀在上方，无论夜晚白天都能有情境的体验，打造出人们流连忘返的饮茶空间。

Space frame extending inside and outside: Specially drawn from the circular window outside the female school, outside the compartment, I saw a painting inside the room. The Grouped tea drinking space use the broken space to separate one person, Valentine's seat, and three-five friends' seats. At the same time, to meet the gathering needs of all ethnic groups, people can easily find the most suitable tea corner in the space. The Starry embellishment dangle chandeliers to embellish the space, creates a unique memorizable drinking experience.

破壳计划扣钉
Coding Life

项目地点：中国陕西 / Location : Shanxi,China
项目面积：80 平方米 / Area : 80 m²
设 计 师：敖静 / Designer : Ao Jane

IAI 设计优胜奖

敖静　Ao Jane

宜空间设计创始人，拥有十余年室内设计从业经验的资深设计师。主导设计项目,包括地产样板间、商业、住宅等等。
她主张设计能够帮助人们更好的理解生活，发现生活中你不曾在意的细节、故事。

Ao Jane , created the easy space design. Leading design projects, senior designer with more than ten years of experience in interior design including real estate showrooms, commercial, residential and more. She advocates that design can help people better understand life, discover details and stories that you didn't care about in life.

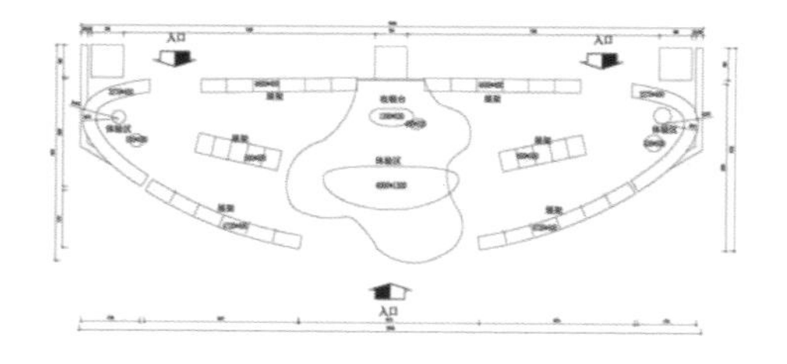

破壳计划，来源于一个创造新生命的理念，蛋壳为了保护，而破壳则为了新生。

80平方米的商业空间，品牌名“Coding life”，中文直译“扣钉”。

店铺外观设计，以钢结构的镂空为基础，与白色蛋壳弧面衔接，以蛋黄木饰面做延伸，三种元素的组合，完整地建立了一个蛋壳的主题造型，这是一个打破重生的新计划，是主题，亦是店招。我们让动线成为了功能区域的划分线，让产品成为动线上的主角，简洁、不失趣味。

破壳，是一个建立、成长，再打破的过程，是店主的创业情怀，也是这个设计的内涵。

Breaking shell program, derived from a concept: breaking the constraints of the eggshell to create a new life.
The brand "Coding life", with its Chinese transliteration "Coding Life".
The exterior design of the shop is based on the skeleton of steel structure, connected with the white curved surface of the egg shell which extends to the egg yolk veneer. This combination of the three elements completely established a egg shell theme which is a plan of breaking and birth.We let the circulation design divide different functional regions, so making the products as the focus on the way customers move through, which makes the circulation design concise, without losing entertaining.
Breaking the egg shell, a process of breaking, establishment and then growth, is not only the entrepreneurial feelings of the shopkeeper, but also the connotation of this design.

依格宝星级发廊
Chengdu IL COLPO Salon

项目地点:中国四川 / Location : Sichuan, China
项目面积:80 平方米/Area : 80 m²
公司名称:北京艾奕空间设计有限公司 / Organization Name : AE ARCHITECTS

IAI 设计优胜奖

AE ARCHITECTS 艾奕设计

依格宝星级发廊（成都）的设计理念将现代、简约、都市生活方式与女性之柔美完美融合。项目本身面临诸多挑战。入口处巨大的柱子将空间一分为二。店铺后方的空间向左延伸，打破了整个空间的连贯性。为了把所有限制因素充分转换成优势，一个将大柱子包含其中的岛台式接待区油然而生。这个设计造型有助于区分接待区与休息区，并且在一个绝佳的位置展示了店铺的品牌标志。

在坐席区的中间，为了划分多个区域，增加储藏空间，我们设计了一个完全被镜面覆盖的隐形盒子。这个集美观和功能于一体的大盒子为染发区提供产品储藏空间。在店铺前端的主要空间里，触感十足的大理石和柔和的色调定义了剪发区。在店铺的后方，我们设置了洗发区、休息室和VIP区。设计所选用的几何元素是一系列层次分明的重叠圆环，表现形式或是空间的大圆柱，或是墙上镂空的置物架，或是天花下的大吊灯。这一系列的安排将移动感诠释得淋漓尽致，进而将店铺的前后方很好地连结在一起。静谧蓝和浅粉色给人一种家一样的亲密感，而最私密的空间则截然相反，是一种高科技、明亮夺目的既视感。

Chengdu IL COLPO Salon combines in its concept the ideas of modern, simplicity, urban life style and soft feminine beauty. The site itself involved several challenges to face.?By the entrance a monumental column divide the space in two. At the back the site moves to the left braking the continuity.
In order to use all this constrains as an advantage we set an island reception desk which include the column in its composition, this structure helps to divide the entrance area from the resting area and also support the logo of the shop.
In the middle of the seating area with the intention to enlarge the number of sections and at the same time have more room for storage we created an invisible box entirely cover by mirror, which serves as hair coloring section and as a storage for products inside.In the perimeter of the main area a touch of marble and soft colors define the cutting hair sections.
At the back of the shop we set the washing hair, restrooms and the vip area.
The overall geometry selected is an overlapped sequence of rings, in the space turn into cylinder, in the wall as a shelve and in the ceiling as a lamp. This sequence arrangement brings the idea of motion and allow us to connect the back of the shop with the shop front.
Calm blue and light pink gives a feeling of home and intimacy while the most private room behave by contrast with a high tech and bright look.

蜕变
Transformation

项目地点：中国江苏 / Location : Jiangsu, China
项目面积：1050 平方米 /Area : 1050 m²
设 计 师：董则锋 / Designer : Dong Zefeng

IAI 设计优胜奖

董则锋 Dong Zefeng

中国建筑装饰协会高级室内建筑师、高级住宅室内设计师，CIID 中国建筑学会室内设计分会会员。1996 年毕业于南京艺术学院装潢设计专业，擅长欧式、新古典、美式、中式风格。

美国 Best of Year Awards 2018 年度最佳设计大奖
亚太设计大赛 - 商业空间 - 钻石奖
亚太设计大赛 - 办公空间 - 至尊奖
华语设计领袖榜 2018 年度 100 位卓越设计人物

Senior Interior Architect, Senior Residential Interior Designer of China Building Decoration Association, member of CIID China Architecture Society Interior Design Branch. In 1996, he graduated from Nanjing Art College with a major in decoration design. He is good at European, Neoclassical, American and Chinese styles.

US Best of Year Awards 2018 Design of the Year Award
Asia Pacific Design Competition - Commercial Space - Diamond Award
Asia Pacific Design Competition - Office Space - Supreme Award
Chinese Design Leaders List of 100 outstanding design characters in 2018

健身会所一楼空间以书吧为主题，书架到顶侧向排灯光，增加空间的整体简洁感受。旋转滑梯寓意蜕变，上宽下窄，从大到小，寓意从胖宽到瘦窄的变化，同时它又是一个龙卷风造型，有一种拔地而起的力量。二楼用赛道的概念将空间分为跑步区、器械区、休息区、操房、瑜伽空间、男女更衣室、造型区，采用金属网格来增加空间的力变与张力。

The first floor space of the fitness club takes the book bar as the theme.
The bookshelves are lined up with lights from the top to the side to increase the overall concise feeling of the space. Rotary slides imply transformation, wide above and narrow below, from big to small, implying changes from fat to thin, at the same time it is a tornado shape, there is a pull-up force.
The second floor space is divided into running area, equipment area, rest area, gymnasium, yoga space, men and women changing clothes, modeling area by the concept of track. The metal mesh is used to increase the force and tension of the space.

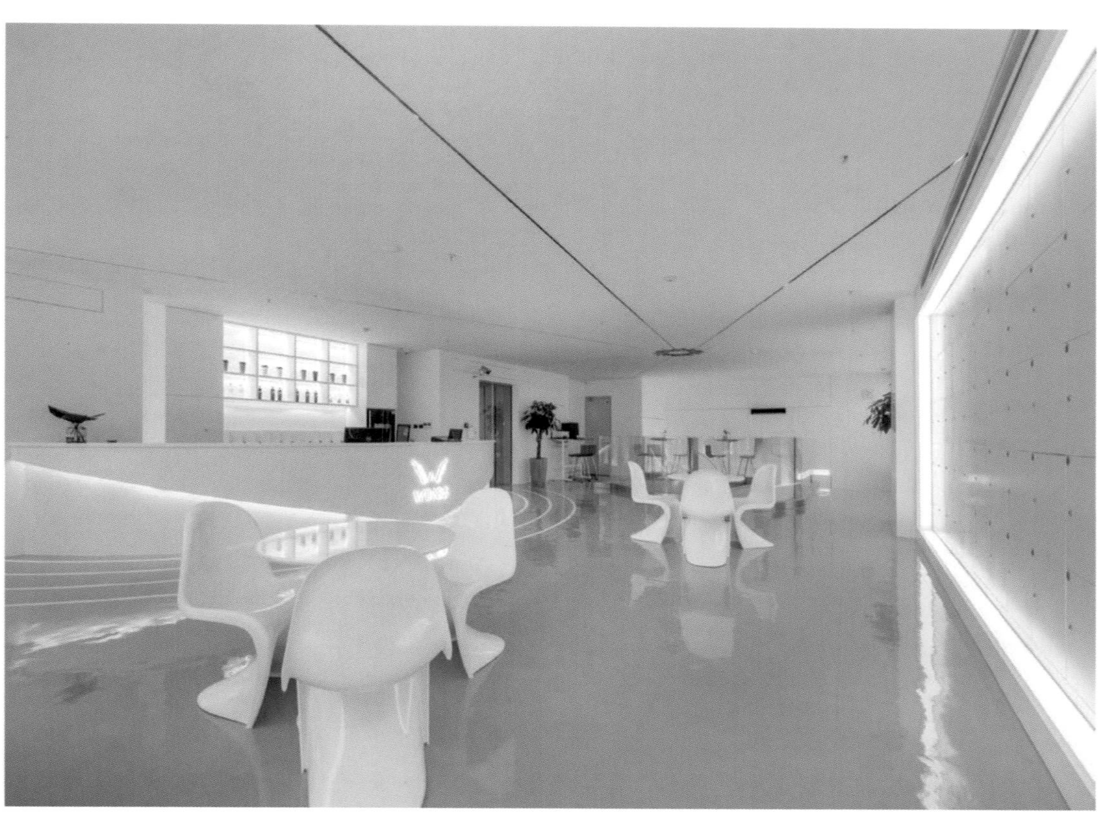

HOIST
HOIST

霍夫曼
HOLFMANN

项目地点：法国 / Location : France
项目面积：200 平方米 / Area : 200 m²
设 计 师：杜柏均 / Designer : Du Bojun

IAI 设计优胜奖

杜柏均 Du Bojun

从业 30 年来，一直秉持着将生活哲学融入个人设计的理念，作品透露着丰富的人文主义精神。受邀北京清华大学研修讲师，并屡次受邀各类设计大赛评委及专业论坛演讲。

自 2004 年来上海创办之后，由杜柏均亲自带领，先后为各大地产集团、品牌合作了许多设计项目，获得美国、英国、意大利、法国、北美等国际大奖，是胡润百富榜上最受青睐的华人设计师之一。

In the past 30 years, he has been adhering to the idea of integrating life philosophy into personal design, and his works reveal rich humanistic spirit. He has also been invited to lecture at Tsinghua University in Beijing, and has been invited to give speeches at various design competitions, judges and professional forums.
Since the founding of Shanghai in 2004, Dubai has personally led many design projects for major real estate groups and brands for more than ten years. It has won international awards such as the United States, Britain, Italy, France and North America. It is also one of the most popular Chinese designers on the Hurun 100 Rich List.

在这个名为“霍夫曼”的展厅里，设计师也为6盏特别的水晶灯建立起了一个如《湮灭》中X区域般神秘而绮丽的场景，但它并非电影所传达的湮灭之地，相反地，你会在这里每个不同场景的位置与角度寻找到各种引人入胜的异域共生。“霍夫曼”依据6盏水晶灯的特性，通过极具设计感的墙体分割成6大不同的功能区域。每一盏灯饰既是主，又为客，成为各自独立空间中指引观者前行的讯息。

In a showroom named "HOLFMANN", the designer also created a mysterious and fancy scene for six special crystal lamps, just like the Zone X in "Annihilation". Unlike the place of annihilation as depicted in the film, you'll find fascinating exotic symbiosis in different locations from different angles.In view of the characteristics of six crystal lamps, "HOLFMANN" is divided into 6 different functional areas by the delicately designed wall. Acting as both a host and a guest, each lamp gives the information to guide the viewer forward in each independent space.

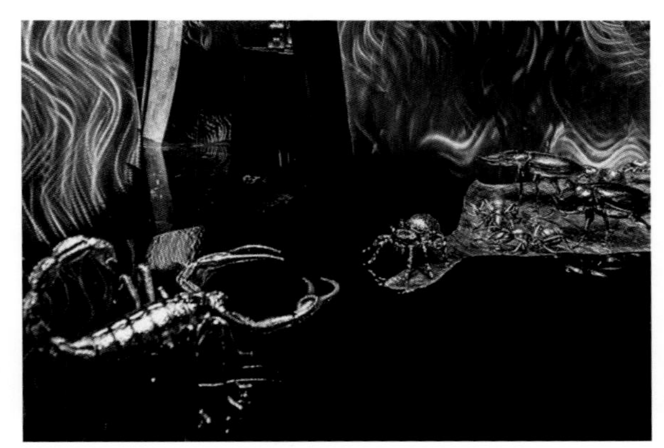

空间中所有的承重结构都被隐藏在了墙体内，手工波形纹理的黄铜饰面与光影形成美妙的流动感，又像是微观世界里的生物肌理，连接着天地，向四处散播着勃勃生机。而墙体缝隙间的镶嵌钻面处理与内置的可变灯带则将广袤宇宙中绚烂的银河藏入其间，蓝绿色大理石与闪亮的水晶钻石强化了灯光的作用，让这里孕育出一股磅礴的能量，蓄势待发。那些自由曲度的墙体在起着区隔空间功能作用的同时，又可视作为空间内独立的艺术品。雕花铜墙面在虚实之间对于步入其间的观赏者而言，可以起到潜意识的位移引导，加之各区域主灯华丽的聚光效果，一条自然而流畅的动线便随之产生。

All bearing structure in the space was hidden in the wall. The brass finish with hand-made corrugated texture forms a wonderful sense of flow with light and shadow, as if the biological texture of the micro world linking the sky and the earth to spread vitality everywhere. The inlay of diamond surface in the cracks of wall and the built-in variable lamp belt conceals the gorgeous galaxy of the vast universe in it. The turquoise marble and sparkling crystal diamonds enhancing the role of lighting breeds a great energy field which is gaining momentum to show. While the free-curving wall functions as a spatial partition, it can also be seen as an independent artwork in space. Between the real and the imaginary, the surface of engraved copper wall provides the subliminal displacement guidance for the viewers who step into the space. Combined with the gorgeous agglomeration light effect of the main lamp of each area, a natural and smooth moving line comes out.

酷博健身俱乐部
Kubo Fitness Club Kunming

项目地点：中国昆明 / Location : Kunming, China
项目面积：3000 平方米 / Area : 3000 m²
设 计 师：方飞 / Designer : Fang Fei

IAI 设计优胜奖

方飞 Fang Fei

2018 红棉中国设计奖 至尊奖
2018 英国 SBID 室内设计大奖
2018 IAI 全球设计大奖 优胜奖
2018 IDS 国际设计先锋榜 金奖提名
2017 第八届中国国际空间大赛 铜奖
2015-2016 中国设计星 全国十二强
2016 CIID 第十九届空间设计大赛 银奖
台湾 2016TAKAO 室内设计大赛 银奖
2016 第四届中国营造空间设计大赛 金奖
第四届 ID+G 金创意奖国际空间实际大赛专业类金奖
Modern Gym Style 中国健身空间设计大赛 金奖

2018 Cotton Tree China Design Award Supreme Award
2018 British SBID Interior Design Award
2018 IAI Global Design Awards Excellence Winners
2018 IDS International Design Pioneer Gold Nomination
2017 Eighth China International Space Competition Bronze Award
2015-2016 China Design Star National Top 12
2016 CIID 19th Space Design Competition Silver Award
Taiwan 2016TAKAO Interior Design Competition Silver Award
2016 The 4th China Space Design Competition Gold Award
The 4th ID+G Gold Creative Awards International Space Reality Competition Professional Gold Award
Modern Gym Style China Fitness Space Design Competition Gold Award

酷博健身俱乐部的空间排列方式大小适宜，呈现出一种难得的秩序美感，这与高级灰、钢琴黑的色调控制有关，与清水混合土、塑钢、水泥砖、钢网相互搭配的材质有关，更与当代年轻人想要的新型运动空间有关。

The space arrangement of Kubo Fitness Club is suitable in size, showing a rare sense of order and beauty. It is related to the control of the hue of the advanced grey and piano black. It is related to the material of mixed soil, plastic steel, cement brick and steel net, which is related to the new sports space that the young people want.

GAIN!

FIGHT

TRX

洵都医疗上海总部
Xundu Health and Beauty Shanghai Headquarters

项目地点：中国上海 / Location : Shanghai, China
项目面积：412 平方米 / Area : 412 m²
设 计 师：赖仕锦 / Designer : Lai Shijin

IAI 设计优胜奖

赖仕锦 Lai Shijin

先后在多家国外设计机构上海分公司担任专案设计师、设计总监。室内设计项目遍及各省市及东南亚等国家，涉及多个领域的室内设计，擅长酒店、办公室、商业及住宅设计。

2012 亚洲室内设计竞赛中国区优胜奖
2012 国际空间设计大赛"艾特奖"入围奖
2016 年度新锐设计师（中国建筑装饰协会颁发）
2017 年度精英设计师（中国建筑装饰协会颁发）

He has worked as a project designer and design director for a number of foreign design agencies in Shanghai branch. Interior design projects in various provinces and cities and Southeast Asia, involving interior design in a variety of areas, specializing in hotel, office, commercial and residential design.

Publishing:
2012 Asian Interior Design Competition China Winner Award
2012 International Space Design Competition "Aite Award" Finalist Award
2016 new designer (issued by China Building Decoration Association)
2017 Elite Designer (China Building Decoration Association)

本案是一个医疗美容机构的办公总部，业主希望办公空间有一部分可以作为接待艺术中心，因此在设计上需要有突出的亮点。

场地的面积有限，为了让空间给顾客留下深刻的第一印象，设计师把前面的空间作为大厅的形式来表现。这个大厅结合了接待、会议、宣传等多种功能，玻璃幕墙的使用，让整个空间更加通透。此外，设计师还利用四根立柱做了一个椭圆造型，并通过一些线条的设计，让空间生动而不呆板。

中间的黑色金属通道衔接了办公区域和大厅，使空间更加连贯。

This case is the office headquarters of a medical and beauty organization. The owner hopes that part of the office space can be used as a reception art center, so there should be prominent highlights in the design.
The area of the site is limited. In order to make the space leave a deep first impression on the customers, the designer takes the space in front as the form of the hall. This hall combines various functions such as reception, meeting, publicity, etc.
The use of glass curtain wall makes the whole space more transparent. In addition, stylist still made an ellipse modelling with 4 stand column, the design that passes a few lines, make the space is vivid and not inflexible.
The black metal passage in the middle connects the office area and the hall, making the space more coherent.

INTERIOR
DESIGN
REVIEW

FIND WAY

顽皮皮具定制俱乐部
Wappy

项目地点：中国广州 / Location : Guangzhou, China
项目面积：275 平方米 / Area : 275 m^2
设 计 师：刘咏诗 / Designer : Liu Yongshi

IAI 设计优胜奖

刘咏诗 Liu Yongshi

2008 年毕业于广州美术学院展示设计，从事商业展示空间 12 年，先后任职于知名外资设计公司，服务过多个国际品牌，拥有丰富多变的设计经验。致力于商业展、空间的创新和艺术呈现的设计工作。2017 年度中国室内设计杰出青年设计师、设计作品曾荣获 IAI 国际最佳设计大奖 Idea-Tops 展示空间设计铜奖等。

Graduated from Guangzhou Academy of Fine Arts in 2008 and has been engaged in commercial display space for 12 years. He has worked in well-known foreign-funded design companies, has served many international brands, and has rich and varied design experience. Committed to the commercial exhibition of space innovation and artistic presentation.

Awards : 2017 Outstanding Young Designers and Design Works of China Interior Design won the IAI International Best Design Award - Dining Space Idea-Tops Exhibition Space Design Bronze Award.

为了更好地传播皮艺的文化精髓，顽皮品牌创始人创立了WP皮具定制俱乐部作为品牌的腹地。空间须要承载传播、分享皮艺创意，也兼负提供个性化定制、奢侈皮具护理的综合功能，同时以艺术的共融模糊三者的风格界限，将皮艺文化的精髓尽情演绎。

The WP brand founder create the customized club as root of WP in order to spread the leather-craft culture soul to each customer. Spreading and sharing the idea of leather-craft, also providing the personal customized and luxurious leather-ware care services are the necessary part of the space. Using art to integrate three of them to deduce the leather-craft culture soul deeply.

慢吧
Slow Bar

项目地点：中国上海 / Location : Shanghai, China
项目面积：30 平方米 / Area : 30 m^2
设 计 师：钱银玲 / Designer : Qian Yinling

IAI 设计优胜奖

钱银玲 Qian Yinling

意大利米兰理工大学国际室内设计管理硕士
上海 TIMESPACE 创始人兼设计总监
上海同济大学室内设计装饰艺术设计学士学位
上海市十大青年优秀设计师
上海市十大青年高端创意人才
APDF 亚太设计师联盟资深会员

Master of Interior Design Management at Politecnico di Milano
Founder and Chief Design Director of TIMESPACE Design
Bachelor of Interior Design and Decorative Arts Design at Shanghai Tongji University
Top-Ten outstanding young interior designers in Shanghai
Top-Ten young high-end creative talents in Shanghai
Senior Member of Asia Pacific Designers Federation

慢吧是一家开在上海静安区高级商业区繁华地段的清吧，设计师特意用内敛含蓄的空间引导顾客的情愫，引发思考与情感融入。

整体设计元素提炼于蒙德里安的抽象直线构成，隐而不发，含而不露，低反差的用色、柔和的灯光、主陈列架流畅回旋的造型，无锋无芒，却营造出没有声音却可静观的安宁。与此同时，当你在桌前享用一杯醇厚变化的鸡尾酒时，你的感官也会随之释放。简化到极致的几何抽象艺术构成，传递出空间与人交流的内在精神。

This is an extremely small business space case. In commercial design, how to highlight the brand's characteristics and temperament in the design of small space is the key to success.
Slow Bar is a clear bar in the bustling area of the high-level business district of Jing'an , Shanghai. The designer deliberately uses introverted and implicit space to guide the customer's feelings, triggering the integration of thinking and emotion.
The overall design element is refined in the pure reproduction of Mondrian's abstract line, hidden but not exposed. The low-contrast color, the soft light, the sleek shape of the main display rack is all hiding the light under a bushel. However, it creates a tranquility that can be quiet without the sound. At the same time, when you enjoy a mellow cocktail at the table, your senses will be released, simplified to the ultimate geometric abstract art composition, and convey the inner spirit of space and people communication.

采用木材、铜、镜面和水泥的肌理，与抽象线条的构成形成良好的对比，将复杂的世界不断地提炼和深化，用艺术的思考呈现出事物内部的秩序与纯粹的精神。

蒙德里安的抽象化与直线构成通过铜与水泥的对比赋予了全新的诠释。半透的窗洞，前堂的转折，隐隐约约看见内吧里调酒师的身影，充满故事感。

Using the texture of wood, copper, mirror and cement, the composition of abstract lines is formed into a good contrast, the complex world is continuously refined and deepened, and the internal order and pure spirit are presented by artistic thinking.

Mondrian's abstraction and linear art composition give a new interpretation through the contrast between copper and cement. You can faintly see the figure of inside Bartender through the semi-transparent window and the turning point of the front hall, full of story.

奢侈品展厅
Luxury Showroom

项目地点：意大利 / Location : Italy
项目面积：400 平方米 / Area : 400 m^2
公司名称 / Designer : Alter Ego Project Group

IAI 设计优胜奖

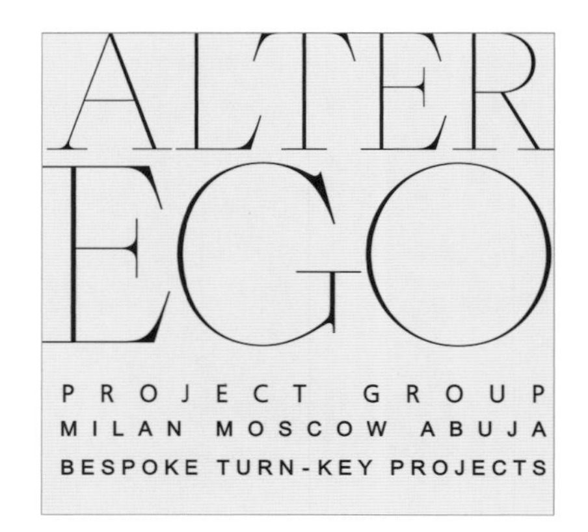

ALTER EGO项目组自2003年成立以来，ALTER EGO项目组始终致力于将指定环境转变为奇迹之地，并提供以下服务范围：定制建筑概念的开发、复杂建筑项目的全面管理、私人住宅豪华内饰的制作、游艇、喷气式飞机和商业地产，设计奢侈品。项目组的重点是定制和独家、材料的选择和交付（装饰材料、照明、家具、装饰物）。

From inception in 2003 ALTER EGO Project Group stays fully committed to transforming designated environments into places of wonder and delivering the following scope of services: development of bespoke architectural concepts, full scale management of complicated construction projects, crafting of deluxe interiors for private residences, yachts, jets and commercial properties, designing of luxury pieces with the emphasis on customization and exclusivity, the selection and delivery of materials (finishing materials, lighting, furniture, décor objects).

ALTER EGO项目组在米兰市中心的18世纪晚期意大利历史宫殿完成了一个设计项目。

设计师的灵感来自其地理位置和历史价值。ALTER EGO为了向过去致敬，引入了新的相关内饰元素，同时尊重现有的历史元素——威尼斯地板具有悠久而辉煌的历史，可追溯到文艺复兴时期，天花板上还有精美的绘画和灰泥。该空间配备了最先进的设备。虽然现代技术的整合不会影响内饰的完整性及其感知。内部风格是新古典主义与现代元素的诗意结合。

针对这个空间，ALTER EGO项目集团的总部注定要反映公司的创新精神，并应对各种社会挑战。最主要的想法是创造一个生活方式空间，让人们可以放心地享受最高端的奢华氛围，同时也是公司客户和合作伙伴的完美交汇点。结果是创意背景社区的时尚场所，这个社交平台汇集了不同国籍和文化背景的建筑师、供应商和客户。

ALTER EGO Project Group completed a design project in an Italian Historical Palazzo of the late 18th century in the very heart of Milan.
The designers were inspired by its location and the historical value. Paying tribute to the past ALTER EGO introduced new relevant interior elements while respecting the existing historical ones - Venetian floors with a long and distinguished history back to the Renaissance period, ceilings with a subtle painting and plasterwork. The space is outfitted with state-of-the-art equipment. Though the modern technologies were integrated without compromising the integrity of the interior and its perception. The style of the interior is a poetic combination of neoclassicism with contemporary elements.

This space, the Head Quarter of ALTER EGO Project Group, was destined to reflect company' s creative spirit and meet various social challenges. The main idea was to create a lifestyle space where people can feel at ease plunging in the atmosphere of the outmost luxury, at the same time the perfect meeting point for company' s clients and partners. The result is a stylish venue for the creative community, the social platform that unites architects, suppliers and clients of various nationalities and cultures.

东亚照明
Light Space

项目地点：中国台湾 / Location : Taiwan, China
项目面积：235 平方米 /Area : 235 m²
公司名称：如沐室内装修有限公司 / Organization Name : Zoom Interior Design

IAI 设计优胜奖

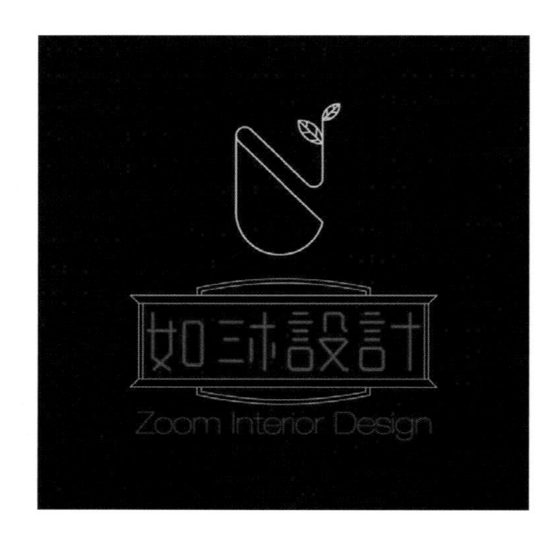

如沐设计 Zoom Interior Design

如沐设计成立于 2014 年，在中国台湾的设计生态中，我们像是从裂缝中极力窜出的新芽，没有充足的养分却又希望能够成长茁壮。空间的定义对于如沐设计是重如泰山，因此不断反复与业主沟通需求是我们最为重视的过程，以人为本是不变的真谛，量身打造业主对于家的憧憬，结合如沐设计师们历年淬炼的生活美学，创造隽永而温润的美感空间，成就每个经典，收藏完美的作品，这是为我们毕生贯彻的设计意志。

Zoom Design was established in 2014.In Taiwan's design ecology, we are like new sprouts from cracks, and we don't have enough nutrients, but we hope to grow and thrive. The definition of space is as important as the Zoom Design , so the constant communication with the owners is the process that we attach the most importance to. The people-oriented is the true meaning of the people, tailoring the owner's embarrassment for the family, combining the years of the designers. Quenching the aesthetics of life, creating a space of beauty and gentleness, and creating a perfect collection of classic works.It is the design will for our life.

企业总部代表了品牌精神，而玄关柜台乃企业予人的第一印象，我们在柜台的设计上为业主提出贴切的创作概念，最终得到令人满意的结果。我们从一位冶铁者在高温环境下专注打造工艺品过程中所激荡出的火花炫目而灿烂，而铁砧正承受着千锤百炼的画面中获得灵感，用以象征企业主成立近半世纪以来仍兢兢业业塑造品牌，力求在渠道与行销予以变革、创新的经营策略。柜台采制作铁艺的工具“铁砧”为概念，同时也与企业标志的首个英文字母“T”相呼应；创作铁艺时必然会产生火花，我们在空间里运用灯条作为指明，而这些火花最终流动至拥抱世界舞台的愿景之上。

As corporate headquarter represents corporate value, the entrance counter is the first impression of the corporate spirit, so it was a conscious design decision to inject the overall design essence into the counter design, before getting such satisfactory result. The inspiration came from the process of ironsmith in symbolizing the successful corporate strategies in the past 40-plus years to ferment its brand and bringing innovative changes to its channel and marketing. The counter was designed with the concept of iron board, with its T-shape resembling the first English alphabet, “T” , of the corporate logo; as iron work produces spark, we use strip lighting to articulate the dynamic quality of sparkling light.

除了展示企业产品的特点与价值，此处也是员工工作的地方，因此我们精研格局分配的合理性、廊道尺度、空间色彩，并赋予多样化的灯光配置，营造舒适、符合人体工学且高效的办公环境。所有部门被统合在同一平面中，以玻璃为隔间的独立主管办公室前方直接对应所属部门，强化了水平与垂直沟通效率。

Apart from exhibiting corporate product characteristics and values, this is also the place where staffs complete their work, therefore, much care has gone into the rationality of layout, corridor dimension, spatial coloring and diversified lighting arrangement so as to create a comfortable, ergonomically friendly and highly productive office environment. All departments have been included in a single space plane, while managers' offices are separated from, yet connected to, their respective staffs in-charge, in strengthening horizontal and vertical communication effectiveness.

Blanko网红咖啡生活馆

Blanko Online Celebrity Cafe

项目地点：中国北京 / Location : Beijing, China
项目面积：100 平方米 /Area : 100 m²
公司名称：领筑智造 / Organization Name : inDeco

IAI 设计优胜奖

领筑智造 inDeco

领筑智造一站式互联网公共空间设计装修服务机构，提供专业空间设计、装修施工、软装配饰、后期配套等服务。始终以"设计感、高质量、透明化"作为核心理念，拥有极具创意的设计团队、专业化的供应链管理团队及精细、高效的施工团队。

inDeco is The One-stop Solution for Office and Public area Design, Engineering, Technology, Management, Construction and Commissioning. Believing in "Edging design, High quality, Transparency" as the core value. With In-house innovative design team, professional supply-chain team and project management team, inDeco continues to thrive for Efficiency, Delicacy, Transparency.

基于Blanko多元素的经营理念，将来福士店划分为不同的功能区，并采用北欧风格的设计概念，用流畅的线条、简洁的色块，将Blanko打造成为这座城市中平静与放松的驿站。

Based on Blanko's multi-element business philosophy, divided into different functional areas, and adopts the Nordic-style design concept. With smooth lines and simple color blocks, Blanko is built into a calm and relaxing station in the city.

M.O.服装展厅
Modern Outfitters

项目地点：中国重庆 / Location : Chongqing, China
项目面积：880 平方米 / Area : 880 m²
设 计 师：尹晨光 / Designer : Yin Chenguang

IAI 设计优胜奖

尹晨光 Yin Chenguang

重庆非间室内设计有限公司(VSD DESIGN)创始人及设计总监，叁仟界艺术设计研究院（重庆）有限公司设计总监，APDF 亚太设计师联盟专业会员。简单、真实、专注面对人、物、事，设计是一门修心的艺术，让您不断发现问题解决问题。

VSD DESIGN Founder and design director Design Director, San Qian Jie Art Design and Research Institute (Chongqing) Co., Ltd.
APDF Asia Pacific Designers Alliance Professional Member. Simple, true, and focused on people, objects, things . Design is a masterpiece of art that allows you to constantly discover problems and solve problems

这套作品里有极简、有复杂；有亲近、有疏离；有童趣、有时尚。有服装展示区（女装、男装、童装）、网红自拍区、试衣区、咖啡书吧、儿童游玩区，真的可以说每一个身处其中的人，在不同的生活状态或情绪里，都能在不同的空间功能区，找到和这个空间属性相契合的点。

The minimalist, detail is in the works; A close, a comb from; Tong qu, fashion. Have clothing exhibit (women's, men's clothing, children's clothing), web celebrity self-time, super comfortable the fitting area, coffee books, children's play area, really can say every body part, can be in different status, or mood of life, can be in different function space, find and the space attribute corresponds to the point.

宝威地毯展厅
Baowei Carpet Exhibition Hall

项目地点：中国台州 / Location : Taizhou, China
项目面积：430 平方米 /Area : 430 m²
设 计 师：尹晓敏、沈丹 / Designer : Yin Xiaomin, Shen Dan

IAI 设计优胜奖

尹晓敏 Yin Xiaomin
沈丹 Shen Dan

浙江后朴设计合伙创始人。
设计不是一种技能，而是捕捉事物本质的感觉能力和洞察能力，并执着于将设计演绎到极致。擅长室内设计，环境设计，设计了包括了精致私宅、精品商业空间、办公空间、高档别墅等设计项目。

Founder of Houpu Design Partnership in Zhejiang Province
Design is not a skill, but the ability to grasp the essence of things, sense and insight, and persevere in deducing the design to the extreme, creating wonderful design works repeatedly. Being speciliziing in interior design and environmental design. Have designed many design projects including delicate private house, fine commercial space, office space and high-grade villas.

本展厅是以围绕全白色素装打造了一个柔和而且纯净的世界，使空间看起来更大，素色的背景下，地毯线条显得柔美，光鲜明亮。简约又不失大方是本方案的主导思想。本设计摒弃了繁琐与奢华，以自然色调为主，以舒适机能为导向，强调回归自然，使顾客在购买中更加轻松舒适。

This exhibition hall has built a soft and pure world around all white and plain clothes, and has made the space look larger. Under the plain background, the carpet appears soft and bright. Simplicity and generosity are the main ideas of this plan. This design abandons fussy and luxurious, mainly focuses on natural colors, takes comfort function as the guide, emphasizes returning to nature, and makes customers more relaxed and comfortable in purchasing.

整个设计在材料选择上多倾向于硬软结合、光挺。选材上也多取舒适、柔性、温馨的材质，而这样的自然情趣正迎合了人们对于自然环境的关心、回归和渴望之情，有一种家的温馨，能够给消费者焕然一新的感觉。此类设计制造了商店的热络气氛，环境提高了顾客的购买情绪。

The whole design tends to be hard and soft and smooth in material selection. The selection of materials is also made of comfortable, flexible and warm materials, and such natural tastes cater to people's concern, return and desire for the natural environment, and has a warm home that can give consumers a new feeling. This kind of product creates a hot atmosphere in the store, and the environment improves customers' buying mood.

蓝光锦蓝府
Blue Light Jinlan Mansion

项目地点：中国徐州 / Location : Xuzhou, China
项目面积：100 平方米 / Area : 100 m²
设 计 师：周汉林、师义 / Designer : Zhou Hanlin , Shi Yi

IAI 设计优胜奖

周汉林 Zhou Hanlin

南通禾墅装饰工程有限公司创始人
擅长领域：酒店、会展展示、办公室内设计
2018 IAI 商业类设计优胜奖
2017 金堂奖
中国室内设计年度评选年度优秀休闲娱乐奖
2015 全球先锋设计（中国）大奖餐厅组别铜奖

Founder of Nantong Heshu Decoration Engineering Co., Ltd.
Areas of expertise: hotels, exhibitions, office design
2018: IAI Design Award Excellence Award
2017: Golden Hall Award China Interior Design Annual Awards Annual Outstanding Entertainment Award
2015: Bronze Award for Global Pioneer Design (China) Grand Prize Restaurant Group

师义 Shi Yi

国家注册一级建造师
注册中级工程师
2015 复旦大学照明研修班进修
2018 IAI 商业类设计优胜奖
2015 中国建筑协会南京室内设计大赛、灵山地铁控制中心方案类一等奖

National Registered Level 1 Architect
Registered intermediate engineer
2015: Fudan University Lighting Training Course
2018: IAI Design Award Excellence Award
2015: China Building Association Nanjing Interior Design Competition The first prize of the Lingshan Metro Control Center

项目位于徐州泉山区淮海天地SOHO售楼处，考虑空间层次合理布局，我们将原始建筑做了相应改造。

The project is located in the SOHO sales office of Huaihai Tiandi, Quanshan District, Xuzhou. Considering the reasonable layout of the space level, various elements allows customers to penetrate into them and accelerate of customers to real estate.

SY.集木

SY.Collecting Wood

项目地点：中国重庆 / Location：Chongqing, China
项目面积：380 平方米 / Area：380 m²
设 计 师：周令 / Designer：Zhou Ling

IAI 设计优胜奖

周令　Zhou Ling

周令装饰设计有限公司设计总监
CIID 中国建筑学会室内设计分会
重庆青年设计师俱乐部 3.0 理事
IAD 国际设计师协会成员
WDC 世界设计联盟成员
APDC 国际设计交流中心资深会员
IAI　亚太设计师联盟成员

Zhou Ling Decoration Design Co., Ltd. Design Director
CIID China Architecture Society Interior Design Branch
Chongqing Young Designers Club 3.0 Director
Member of the IAD International Designers Association
WDC World Design Alliance member
Senior Member of APDC International Design Exchange Center
IAI Asia Pacific Designers Alliance member

体验者在享受高品质牛肉的同时也能感受到空间的魔力，每一个室内的建构都是一个独立的建筑体，而每一个建筑体都有最合适自己的空间形态。

以弧形作为整体设计的主线，而与生鲜和牛所匹配的材料当然也会是最自然的木材与石皮，所以用两段弧形的线条作为整体空间设计的引导线，首先向体验者讲述牛肉的品质和分类，然后讲到调理和食材的配比，最后引导体验者进入到“餐厅”欣赏厨师对整个牛排的烹饪过程。

The experiencer can feel the magic of space while enjoying the high quality beef. Every building in the room is an independent building, and each building has the most suitable space form.

The arc is the main line of the whole design, and the material matching the fresh and the cattle will certainly be the most natural natural wood and stone skin. So the two curve lines are used as the guide line for the whole space design. First, the quality and classification of the beef are told to the experiencer, and then the ratio of conditioning and food is told, and the final guide is guided. The experiencer goes to the "restaurant" to appreciate the cook's cooking process for the whole steak.

墨客体验馆
MoKe Experience Pavilion

项目地点：中国北京 / Location : Beijing, China
项目面积：28 平方米 / Area : 28 m²
设 计 师：周游 / Designer : Zhou You

IAI 设计优胜奖

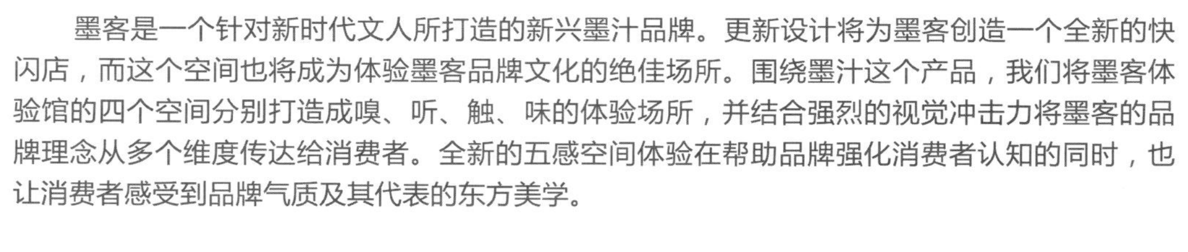

墨客是一个针对新时代文人所打造的新兴墨汁品牌。更新设计将为墨客创造一个全新的快闪店，而这个空间也将成为体验墨客品牌文化的绝佳场所。围绕墨汁这个产品，我们将墨客体验馆的四个空间分别打造成嗅、听、触、味的体验场所，并结合强烈的视觉冲击力将墨客的品牌理念从多个维度传达给消费者。全新的五感空间体验在帮助品牌强化消费者认知的同时，也让消费者感受到品牌气质及其代表的东方美学。

MoKe is a newly developed ink brand for the new generation of literati. This case is a brand new flash shop created for MoKe users, it will become an excellent place to experience MoKe brand culture. Around the product of ink, the designer makes the four spaces of the MoKo experience hall to be a place of smell, hearing, touch and taste, and conveys the brand concept from many dimensions to consumers with a strong visual impact. The brand-new five-sense space experience not only helps the brand to strengthen the consumer's cognition, but also makes the consumer feel the brand temperament and the oriental aesthetics on behalf of the brand.

周游 Zhou You

更新设计创始人、首席执行官
中央美术学院建筑学学士
纽约普瑞特艺术学院设计管理硕士

Update Design Founder & CEO
Bachelor of Architecture,Central Academy of Fine Arts.
Master of Design Management, Pratt Institute, New York.

侨信云起销售中心
Qiao Xin Clouds Up Sales Centre

项目地点：中国广东 / Location : Guangdong, China
项目面积：2300 平方米 / Area : 2300 m²
设 计 师：朱海博、钟行建 / Designer : Asoka Chu , Ken Chung

IAI 设计优胜奖

朱海博 Asoka Chu

知名室内建筑师
深圳市室内设计师协会副秘书长
意大利米兰理工大学国际室内设计硕士
深圳市故事空间设计有限公司创始人
壹同创意集团创始人

Famous Interior Architect
Deputy secretary general of Shenzhen association of interior designs
Master of International Interior Design, Politecnico di Milano, Milan, Italy.
Founder of Shenzhen Storybox Design Co., Ltd.
Founder of TO GET THE Creative Group

钟行建 Ken Chung

壹同创意集团创始人
KEN 设计事务所创始人
意大利米兰理工大学国际室内设计硕士

Founder of TO GET THE Creative Group
Founder of KEN Design Studio
Master of International Interior Design, Politecnico di Milano, Milan, Italy.

金地自在城售楼处

Seaport Golden Land Free City Sales Office

项目地点：中国海南 / Location :Hainan, China
项目面积：1000 平方米 / Area : 1000 m²
设 计 师：邹洪博 / Designer : Born Zou

IAI 设计优胜奖

邹洪博 Born Zou

曾为多个高端住宅、酒店项目担纲主创设计，拥有十多年的室内设计经验。作品提倡人与空间的互动，追求创新，强调品味、舒适的空间环境。他希望可以通过多年积累的设计经验，为国内的客户提供国际化的设计理念及优质的设计服务，给室内设计行业带入一股新的能量。

Born Zou has worked as a master designer for several high-end residential and hotel projects and has more than 10 years of interior design experience. The work promotes the interaction between people and space, pursues innovation, and emphasizes the taste and comfort of the space environment. He hopes that through years of accumulated design experience, he will provide domestic customers with international design concepts and high-quality design services, bringing a new energy to the interior design industry.

“大海胸襟，椰树风骨，三角梅品格”是海口城市精神，与都市的时尚感相约，建筑、自然、水景、光的完美结合也与海口本身的石山火山群地理条件相得益彰；海口风貌诱人，有“四时常花，长夏无冬”之说，运用最自然的原木与极具现代感的点、线、面相结合，自然生态与都市感完美诠释，点缀海与韵、韵与彩。

"The sea mind, the coconut trees of character, bougainvillea character" is the spirit of Haikou city, and urban sense of fashion, architecture, natural and the perfect combination of waterscape, light and stone mountain of Haikou itself also volcanoes geographical conditions bring out the best in each other; Haikou looks attractive, has a "four seasons flower, long summer without winter," said, integratingthe most natural logs and dotted line that has contemporary feeling extremely field, natural ecology and urban perfect feeling, dotted with sea rhyme, rhyme and colour.

Hotel and Resraurant Space

室内 —— 酒店/餐饮空间

洞穴
Cave

项目地点 : 中国广州 / Location : Guangzhou, China
项目面积 : 1166平方米 /Area : 1166 m²
设计师 : 黄永才 / Designer : Ray Wong

IAI 最佳设计大奖

黄永才 Ray Wong

共和都市创始人、创意总监、中国设计鬼才，致力于以革新的精神为社会提供多样性的设计产品，以创意、富有想象的创造力设计驱动更多的品牌价值，作品涵盖建筑、顶级 CLUB, 餐饮办公场所、私宅等，创新大胆的设计为空间注入新的活力，作品获得多项顶级全球赛事奖项。

Republican City Founder & Creative Director & Chinese Design Genius,
Committed to providing a diverse range of design products to the society in an innovative spirit.
Designed to drive more brand value with creative, imaginative creativity, including architecture, top-level CLUB, dining and office space, private homes, etc.
Innovative and bold design injects new vitality into the space.
Works won several top global competition awards

“Cave”空间来自“洞穴”的概念，源于原始山洞里的篝火和温暖，处理人与空间，空间与空间之间的界面关系。大家簇拥而坐，有歌声有酒，单纯而美好。

洞穴的创造，结合球体和墙壁，是整个空间的基本界面。设计使人与空间，空间与空间的关系既独立又相互渗透。

The idea of CAVE project stems from ancient cave, from its interfacial relationship between gathered-twigs fire and warmth handling the interface between people and space, space and space.While mentioning ancient cave, the pure and good scene,like people clustering around fire, singing and drinking,will then occur in our mind.

Designer would like to utilize the concept of cave on the interfacial relationship between people and space, space and space.The creation, the cave-holes, combining sphere and wall, is used as the fundamental interface for the whole space . Design makes the relationship between people and space, space and space independent and infiltrated.

这些“洞口”在外观上形成了特点鲜明的建筑的门窗，半圆的门窗的构造在众多的横平竖直的建筑环境中别具一格。透过外观的这些半圆形的门窗，无论哪个角度我们都可以看到cave club的吧台。隐约的灯光、动人的音乐，都勾起了人们进去一探究竟的欲望。内部的空间是由不同的洞口围合而成的半开放的包间和开放的大厅，空间独立又暧昧。这些空间不再是传统的间隔，空间所有的元素造型都是灵动的、自由的。空间的半开放使得整个空间的所有时间、事件、人物、视觉、嗅觉、听觉都感觉融合叠加在一起了。体验感是开放的，却又是独立的；使得人与空间、空间与空间之间既独立又相互渗透。

These semi-round door and windows make it stand out the boxy buildings. Through any exterior door and windows, the bar table of CAVE CLUB can be seen at any angle. After steping inside, people will find it is a futher experience than standing outside.After steping inside, people will find it is a futher experience than standing outside. The space is independent but also with obscure boundary, semi-open booths and open-hall are all shaped by variety-size holes.

These spaces are no longer the traditional spacing, space all the elements of the modelling is clever and free. The semi-open space makes the whole space all the time, events, characters, vision, smell, hearing all feel fusion superposition together.Experience is open, but independent; So that people and space, space and space between the independent and mutual penetration.

泉上海鲜粥
SEAFOOD CONGEE

项目地点：中国贵阳 / Location : Guiyang, China
项目面积：500 平方米 /Area : 500 m²
设 计 师：钱银玲 / Designer : Qian Yinling

IAI 最佳设计大奖

钱银玲 Qian Yinling

意大利米兰理工大学国际室内设计管理硕士
上海 TIMESPACE 创始人兼设计总监
上海同济大学室内设计装饰艺术设计学士学位
上海市十大青年优秀设计师
上海市十大青年高端创意人才
APDF 亚太设计师联盟资深会员

Master of Interior Design Management at Politecnico di Milano
Master of International Interior Design Management founder and Chief Design Director of TIMESPACE Design
Bachelor of Interior Design and Decorative Arts Design at Shanghai Tongji University
Top-Ten outstanding young interior designers in Shanghai
Top-Ten young high-end creative talents in Shanghai
Senior member of APDF Asia Pacific Designers Federation

整个空间围绕自然朴素的主题展开，将工业风与艺术相融合，紧扣言粥记的精髓。入口极具工业化的净水装置与吧台相结合，带给人们独特的视觉冲击的同时也在诉说其熬粥水质的用心与健康之处。镂空钢板的一字一句承载了泉上的坚持与灵魂。就餐空间的不同形式组合搭配，既有变化又有效统一；垂直空间互借的下沉空间与夹层空间巧妙形成对比，丰富空间及就餐体验。

he whole space revolves around the heme of natural simplicity, integrat- g the industrial style with art, closely eeping up with the essence of Yan- houji. The entrance is equipped with he industrial water purification evice and the bar, giving people a nique visual impact while also telling he heart and health of the water uality of making porridge. Every ord on the hollow steel plate bears he persistence and soul of the Spring'. Different combination and ollocation of dining space is hangeable but also unified. The hared sinking space of the vertical pace is ingeniously contrasted with he mezzanine space, enriching the pace and dining experience.

粥
SEAFOOD CONGEE

顺应原有建筑结构语言的镂空钢板处理，玻璃幕墙巧妙营造空间灯光的明暗处理，点睛主题的人物油画由艺术造诣极高的艺术家精心创作，更将空间质感高度提升。朴素的水泥、原木、钢板、纱幔充分融入，刚柔并济地诉说着空间语言。让你在整个就餐过程中从空间氛围到味蕾都留下深刻记忆。

he hollowed-out steel plate in accordance with the original architectural language, the shading of the space lighting cleverly created by the glass curtain wall, and the eye-catching theme oil paintings produced by artists with high artistic accomplishments, make the space texture highly enhanced. Simple cement, logs, steel plates, and gauze are fully integrated, speaking the space language in a firm and flexible way.Let you have a deep memory of the space atmosphere and taste buds throughout the meal...

客语餐厅
HAKKA YU

项目地点：中国深圳 / Location : Shenzhen, China
项目面积：500 平方米 /Area : 500 m²
设 计 师：王锟 / Designer : Kevin Wang

IAI 最佳设计大奖

王锟 Kevin Wang

国内知名空间设计大师，艺鼎设计创始人兼首席设计师，坚持完美有趣的设计。凭借敏锐的商业视角和深厚的美学素养，引导客户创造全新的消费体验，作品多次荣获新加坡 INSIDE 年度最佳餐饮空间、英国 WINA 年度最佳餐饮空间、艾特奖等行业权威奖项的认可。

A well-known domestic space design master, founder and chief designer of Yiding Design, insists on perfect and interesting design. With a keen business perspective and profound aesthetics, it guides customers to create a new consumer experience. The work has won many awards such as Singapore INSIDE's annual best dining space, UK WINA's annual best dining space, and Aite Award.

作为一家客家餐厅，艺鼎以历史悠久的客家文化为灵感来源，在空间设计中营造家的氛围。本案以“村落”为主题，运用客家传统建筑的空间形式，并从客家围屋中提炼老门窗和屋檐等元素作为载体，在现代东方的语境下，将其打造成一个极具东方文化又充满现代时尚感的空间，让消费者在用餐的同时也能感受到古朴的客家文化氛围。

入口处土楼建筑的置入，完美诠释了空间主题。因原建筑整体空间纵深较长，设计师在思考整体空间布局时，通过对“村落”建筑布局的借鉴，将各个功能分布于房屋和院子中，错落有致，让空间更具层次感和节奏感。

As a Hakka restaurant, Yiding Design is inspired by the historic Hakka culture to create a homely atmosphere in the space design. As for the case, the theme is “village” . In addition, by applying the space form of traditional Hakka architecture and extracting various elements including ancient doors and windows and eaves from Hakka round houses as a carrier, it is transformed into a space featuring strong sense of oriental culture and modern fashion in the context of modern orient, allowing consumers to experience the quaint Hakka culture while dining.

The placement of the earth building at the entrance perfectly reflects the theme of the space. Since the overall space of the original building is long and deep, the designer has referred to the layout of the “village” while considering the overall spatial arrangement and distributed various functions in houses and yards, in which way, the space is well-proportioned and appears with stronger sense of layer and rhythm.

山野之根

走进客语，迎面是一条风景怡然的村落走道，引人入胜。包间位于道路的过渡区，区域整体抬高做了上层式设计，私密性和独立性也更完整。卡座区的吊顶设计则是提炼了传统屋檐的建筑形式，在掩盖上层杂乱天花的同时，也加深了房屋的形式感。

设计师在表现一种现代建筑体块关系的同时，采用新旧材料碰撞的手法，让空间富有活力。餐厅明档的设计选用玻璃隔断和极具古韵的老门窗作结合，既保留客家文化的历史痕迹，又彰显了时代气息。

客语灯具的设计借鉴了孔明灯的形态，营造了一种逢年过节阖家团圆、齐放孔明灯的场景。而在灯罩上酷似圆月的红色吊顶，则以月亮为灵感，呼应了孔明灯的团圆景象，温馨而有趣。洗手池设计在走道区，光线柔和，在绿植和烟雾缭绕的干冰衬托下，若隐若现，营造了独特的景观效果。此外，空间大部分运用间接光，营造如同在家中用餐的轻松氛围，完美诠释了“家”的概念。

Entering HAKKA YU, you will see a village-style walkway in front of you combined with beautiful landscape pattern design, which is very fascinating. Private rooms are located in the transition zone of the walkway, and the upper layer design is applied for the overall elevation of the area, which will guarantee and better the privacy and independence. The ceiling design in common seats area is actually inspired by the architectural form of traditional eaves. As a matter of fact, it deepens the sense of form of the house while covering the messy ceiling in the upper layer.

While presenting a modern building block relationship, the designer also makes full use of the collision of old and new materials to make the space vibrant. Moreover, as for the transparent shield design of the restaurant, glass partitions and old doors and windows with ancient charms are selected in use, which will not only reserve various history elements of Hakka culture, but alsoreflect the atmosphere of the times.

The design of lamps in HAKKA YU has also drawn on the shape of Kongming Lantern, creating a scene of the reunion of the family in holidays and festivals while releasing Kongming Lantern. Meanwhile, the red ceiling design on the lampshade with the very image of the full moon is inspired by the moon, and complements the reunion atmosphere created by Kongming Lantern, which is cozy, warm and interesting. The hand washing sink is arranged in the walkway area with soft light, and against the background combining green plants and dry ice smoke, it is partly hidden and partly visible, creating a unique landscape effect. In addition, most of the space have used the indirect light to create a relaxed atmosphere which will make customers feel like dining at home, and perfectly reflect the concept of “home” .

阳朔花梦间酒店

Yangshuo Blossom Dreams Hotel

项目地点：中国广西 / Location : Guangxi, China
项目面积：2800 平方米 / Area : 2800 m^2
设 计 师：蒋晓林 / Designer : Jiang Xiaolin

IAI 最佳设计大奖

蒋晓林 Jiang Xiaolin

共向设计创始人创意总监，毕业于中央美术学院建筑系，新锐空间建筑师。
多年室内设计领域探索，致力于东方传统美学精神与当代设计手法的实践，开拓了集地产精品酒店，商业空间，艺术装置多领域的设计实现。

Founder / Creative Director, Co-Direction Design
With educational background in Architecture School of Central Academy of Fine Arts, Jiang Xiaolin has been engaged in interior design practices for several years. As an interior architect, he sticks to the combination of traditional oriental aesthetics and modern design approaches, and so far has completed a variety of exceptional design projects, including real estate clubs and show flats, boutique hotels, commercial spaces and art installations, etc.

设计师由建筑开始着手，从景观到室内/软装，进行了全面的设计规划与改造。建筑外立面采用白色的肌理涂料搭配韵律的木纹格栅，形成协调的序列性。庭院借鉴中式园林的理念，院落用3米高的竹林进行围合，塑造出礼仪层次，确保了庭院的独立性与私密性。大堂采用对称的设计手法，传达出新中式的秩序美学。木质材料打造的书吧、茶室，呈现出沉静的气韵，镂空屏风与光影相映成趣，散发着素雅的东方气质。设计师更将中国传统艺术的精髓留白融入室内设计中，拓宽空间层次的同时更具艺术氛围。酒店共26间客房，设计师提取出具有代表性的地域色彩应用在客房区域，创造不同的情境体验，使空间更具自然的灵动与生机。大面积的落地窗将自然景观引入室内，临窗而立，将景致尽收眼底。顶层还设置了一处观景庭院，放眼远眺，一望无际的山景，缭绕在云雾间。

The designers started from the building to make a comprehensive design planning and transformation from the landscape to the interior and decoration. Building facade adopted white texture coating with rhythmic wood grain grille, forming a coordinated sequence.Borrowing from the concept of Chinese garden, the courtyard was surrounded by bamboo forest of three meters high, shaping levels and ensuring independence and privacy of the courtyard .Symmetrical design techniques is used for the lobby to convey the order aesthetics of new Chinese style.Elegant and low-key wooden materials were used to create a book bar and a tea room, showing a quiet artistic conception. Blankness Of, the essense of Chinese traditional art was integrated in the space, while riching the level of space, it created an artistic atmosphere.The hotel has a total of twenty-six guest rooms,designers extracted and applied representative regional colors for the guest room area to create different situational experiences, giving the space more natural spirituality and vitality. Large French windows bring the natural landscape into the interior space, achieving a panoramic view of the scenery by the window.On the top floor of the hotel there is a viewing courtyard, where endless mountain scenery in the clouds can be viewed.

花梦间酒店 BLOSSOM DREAMS HOTEL

鸡尾酒吧
Deus Cocktail Bar

项目地点：中国上海 / Location : Shanghai, China
项目面积：160 平方米 /Area : 160 m²
设 计 师：朗耀 / Designer : Jeorg

朗耀 Jeorg

朗耀师从 Wolf D. Prix 教授（蓝天事务所）并取得了硕士学位，并且朗耀在阿姆斯特丹和上海的 UNStudio 也工作多年。2013 年，朗耀成立了他自己的工作室 LOFE。2013 年，朗耀与 UCLA 的 3M 未来实验室合作，建造了第一栋 3D 打印房屋。这也是他从 2009 年就开始教授的内容。他在同济大学、玻利维亚的天主大学、复旦大学都教过课，做过讲座和演讲嘉宾。

Joerg graduated from the master-class of Prof. Wolf D. Prix at the university of applied arts in Vienna (Coop Himmelblau) and worked for many years as an associate at UNStudio in Amsterdam and in Shanghai. In 2013 Joerg founded his own office LOFE Studio. Joerg was building the first 3d printed house with the 3M futurelab by UCLA, a format he was teaching since 2009. Other teaching positions include the Tongji University and the Catholic University of Bolivia, Fudan University and various guest crits and lectures.

DEUS是一个舒适的可供人们聊天和休闲的酒吧，它是俱乐部酒吧概念的融合以及升华。以纯粹、认可和感知的结合为设计理念，遵循酒吧的名字“DEUS”， 也就是拉丁语中“上帝”的意思。酒吧的功能元素包括一张8米长的吧桌、一个带16个扬声器以及环绕音响系统的DJ区、一个带沙发的抬高的休息区（包括内置的隐藏存储空间），位于3米高的隐藏框架门后面的一个铜包层的立方体，其中包含了3个男女通用的卫生间和一个盥洗区域，以及从酒吧可以通过隐形门进入的储藏空间。此酒吧的工作范围包括从概念设计到现场施工的监督，以及照明和标志的概念设计。

DEUS is the fusion of clubby bar which doubles up to be a cozy place to chat and sit in cozy chairs. The design concept is a combination of purity, sanction and sensation - following the bar’s name “Deus” from Latin translated “god”. The functional elements include a 8m long cobar table, a dj area with a 16 speaker surround audio system, an elevated lounge area with sofas including built in hidden storage space, and a copper cladded cube which hosts behind a 3m high hidden frame door leading the unisex bathroom including 3 toilets and a washroom, and there is also a storage space accessible from the bar through the invisible door. The work scope included concept design to site supervision as well as concept design for lighting and logo.

美的置业汽车营地66号公路餐厅
"Midea DiChan·HighWay66"Restaurant

项目地点：中国广州 / Location : Guangzhou, China
项目面积：180 平方米 / Area : 180 m²
设 计 师：陈正茂 / Designer : Chen Zhengmao

IAI 设计优胜奖

陈正茂 Chen Zhengmao

设计执行董事、总经理，2007 年毕业于广州大学建筑设计专业，2007 至 2015 年 4 月就职国内大型设计公司，2015 年 5 月加盟广州达艺装饰有限公司，并创立子品牌广州古德室内设计有限公司，（GoodDesignCrew）。有深厚的建筑及装饰设计基础，对整体项目有着完善的控制能力和出色的领导才能。

Design executive director, general manager,Graduated from Guangzhou University in 2007 with a major in architectural design.In April 2007-2015, he worked for a large domestic design company.Joined Guangzhou Dayi Decoration Co., Ltd. in May 2015 And founded sub-brand Guangzhou Good Interior Design Co., Ltd.(GoodDesignCrew).
With a deep foundation in architectural and decorative design,Complete control over the overall project And excellent leadership skills.

设计师陈正茂先生期望能将鹭湖的水元素与树林及禅意的概念运用到项目中去，让度假的宾客在舒适优雅的空间里享用美食的同时能感受到文化的氛围。设计师运用现代的手法去书写传统建筑的基本框架结构，大量的木框朴实地展现建筑结构美学，另外用几何阵列的方式呈现鹭湖的水元素，天花的吊灯同样运用波浪的形式做阵列，形成多层次的视觉感受。

室内餐桌，设计师选用了比较能体现木头自然肌理的锯痕木材作为饰面，与座椅的现代感互相碰撞，产生了丰富的质感对比，局部装饰的红色樱桃雕塑成为绿意盎然的环境中点睛的一笔，使空间轻松有趣。

Designer Amus hopes to bring to the project the water elements of the Heron Lake and the concept of the forest and Zen, so that holiday guests can enjoy food in a comfortable and elegant space while feeling the cultural atmosphere. The designer uses the modern method to write the basic frame structure of traditional architecture, a large number of wooden ripples, simple architectural aesthetic, in addition to the geometric array of water elements of heron lake, ceiling chandelier also use the form of waves to make array, forming multi-level visual sense.

Interior table, the designer chose to reflect the wood natural texture of the saw-cutwood as a surface, and the modern sense of the seat collided with each other, resulting in rich texture contrast, the partial decoration of the red cherry sculpture into a green environment in a bright point, making the space easy and interesting.

醇酒吧
Pure WH

项目地点：中国武汉 / Location : Wuhan, China
项目面积：150 平方米 / Area : 150 m²
设 计 师：洪逸安 / Designer : Hong Yian

洪逸安 Hong Yian

中原大学设计学院博士生
中原大学室内设计硕士
中原大学室内设计学士

PhD student of Zhongyuan University School of Design
Master of Interior Design, Zhongyuan University
Bachelor of Interior Design, Zhongyuan University
2014 Gold Award-Interior Design Award Space Furniture-TID Award

位于武汉宁静老社区里的一家鸡尾酒吧的室内，设计概念是捕捉老武汉市井风貌印象进行发想。在铜框架的空间系统中规划多样的座椅与场景装置来围塑城市场所气质，老主顾们每天背负着不同的故事来到这个有故事的酒吧。

有意无意地在这个有故事的地方与大家分享着人生百态的各种故事，灯光随着夜晚的变化越显浪漫与舒适，热情的灯光氛围伴随着一个生日的派对正在上演，一个温暖而具有城市时尚生活格调的高端交谊场所就此展开。

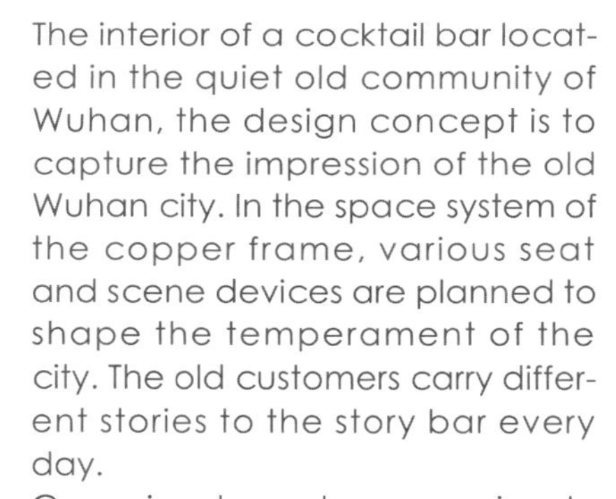

The interior of a cocktail bar located in the quiet old community of Wuhan, the design concept is to capture the impression of the old Wuhan city. In the space system of the copper frame, various seat and scene devices are planned to shape the temperament of the city. The old customers carry different stories to the story bar every day.

Consciously and unconsciously sharing stories of life in this story-filled place. The lighting is more romantic and comfortable with the changes of the night. The warm lighting atmosphere is accompanied by a birthday party, a warm and a high-end social gathering place of urban fashion life style is launched.

榕意中国菜馆
EASE Chinese Cuisine

项目地点：中国广州 / Location : Guangzhou, China
项目面积：500 平方米 / Area : 500 m²
设 计 师：刘咏诗 / Designer : Liu Yongshi

IAI 设计优胜奖

刘咏诗 Liu Yongshi

2008年毕业于广州美术学院展示设计，从事商业展示空间12年，先后任职于知名外资设计公司，服务过多个国际品牌，拥有丰富多变的设计经验。致力于商业展于空间的创新和艺术呈现的设计工作。
本人曾获：2017年度中国室内设计杰出青年设计师、设计作品曾荣获IAI国际最佳设计大奖、Idea-Tops展示空间设计铜奖等。

She graduated from Guangzhou Academy of Fine Arts in 2008 and has been engaged in commercial display space for 12 years. she has worked in well-known foreign-funded design companies, has served many international brands, and has rich and varied design experience. Committed to the commercial exhibition of space innovation and artistic presentation. she has have won: 2017 Outstanding Young Designers and Design Works of China Interior Design
The IAI International Best Design Award - Dining Space, Idea-Tops Exhibition Space Design Bronze Award.

我们以榕意之名，打造“池塘边，榕树下，时光里”的空间氛围。团队经历多次的造型尝试，只为了能以完美的流线打造出榕树气根轻摆的意境。随着气根在墙身的延展，一抹隐隐的榕树剪影落在餐厅尽头。以内敛的形式呈现榕意品牌理念的所在。在2018年榕意荣登《米其林·必比登推介餐厅》，让更多消费者能感受到榕意带来的视觉与味蕾的多重享受。

With EASE as the name, Creating Space Atmosphere as “Along the pond, lean on the banyan, Intoxicated with time.” Our team experienced several times of model creating only focus on showing the perfect artistic conception of banyan with the breeze’ s aerial root. The brand concept of EASE will send to each customer gently by subtle shadow of banyan.

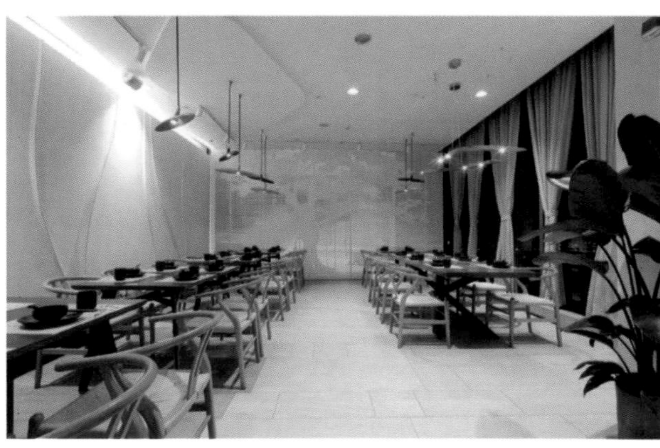

104咖啡馆
104 Cafe

项目地点：中国台湾 / Location : Taiwan, China
项目面积：265 平方米 / Area : 265 m²
设 计 师：吕沛杰 / Designer : Lü Peijie

吕沛杰 Lü Peijie

居间国际室内装修设计执行总监，设计师的立场非常单纯就是提供“观念和方法”，“观念”就是提倡一种生活态度与生活方式，“方法”是以设计为本位。“核心价值”就是做对的事。基于以人为本的设计哲学与初衷来因应每个案件。创造让空间、自然与生活行为环环相扣，让人们适得其所。

PEI CHIEH LV, Execution Director of the InSpace Interior Design Ltd. The standpoint of PEI-CHIEH is very simple, one of which provides “Concept and Strategy” the "Concept" advocates improving the living attitude and lifestyle. The "Strategy" is basing designs on original ideas. The core value is doing things right for the result to fulfill its value. Base on human-oriented design philosophy as the original intention to respond to each project. They design the space, where the space, nature, and living behavior are closely linked and appropriate.

104咖啡馆，不仅仅是一家咖啡馆。

它可以作为一个吃饭、喝咖啡、聚会、小组工作的地方，促使员工进行更多的互动，激发新的想法并促进合作。它的目的是成为一个多功能的地方，而不仅仅是设计师所说的适当的“自助餐厅”或其他“咖啡馆”。

乍一看，弯曲的天花板引导眼睛聚焦到一个空间，该空间与后面的栅栏对齐。它延伸了这个区域，象征着生命的兴衰。为了使这些匹配完美，设计师将品牌颜色扩展到三角形木栅栏。创造一个神秘的私人部分，同时也塑造一个不同层次的空间。

红色表达了激情和警告。两者都巧妙地布置在墙壁中，以便划分空间的不同服务部分，然后进一步扩展尺寸并在空间内创建不同的层次。

除了餐厅的形式和风格，更具挑战性的部分是传达多功能空间的精神。

104 cafe, is not just another cafe.
It holds as a place for eating, coffee breaking, meeting, group working, spurring employees to interact more, sparking fresh ideas and boosting collaborations. It holds the purpose of being a multi-functional place, not just a proper "Cafeteria" or another "Cafe", as stated by the designer.
At first sight, the curved ceiling leads the eyes to focus into a space that aligns into a fence at the back. It extends the area, symbolizing life’ s rise and fall. To make these match perfectly, the designer had extended the branding colours into the triangular wooden fence. Creating a mysterious private section, meanwhile also fashioning a space with different layers.
The red color expresses passion and warning. Both are arranged in the wall cleverly in order to divide the different service sections of the space, which then further extends the dimension and creates different layers within space.
Other then the form and style of the restaurant, the more challenging part was to convey the spirit of the multi-functional space.

芳芳法餐厅
Fang Fang French Restaurant

项目地点：中国苏州 / Location : Suzhou, China
项目面积：495 平方米 /Area : 495 m²
公司名称：拿云室内设计有限公司 /Organization Name : Nayun Design

IAI 设计优胜奖

拿雲設計
NAYUN DESIGN

南京拿云室内设计有限公司成立于 2013 年，公司成立伊始，即致力于商业空间设计的探索与实践，核心的设计团队有着不同的教育和实践经验，使作品呈现出更加丰富多元的创意。设计领域包括商业空间设计、办公空间设计、商业地产等。
作为一个有着丰富经验的设计团队，拿云设计有着自己的设计风格与态度。用创新的激情在业界赢得了口碑和荣誉。

Nanjing Nayun Interior Design Co., Ltd. was founded in 2013, which is devoted to the exploration and practice of commercial space design. The core design team has different education and practical experience, which makes the works show more rich and diverse creativity. Design areas include commercial space design, office space design, commercial real estate, etc. As a design team with rich experience, Nayun Design has its own design style and attitude. With the passion of innovation, we have won praise and honor in the industry.

L'Arome芳芳是以推广“bistronomy”为主线的新式酒馆美食品牌，最新的这家坐落于苏州新地标苏州中心南区三楼，这里是众多时尚年轻人的聚集地，也是都市先锋潮流概念的集中展示区。拿云设计承接了本次的店面设计，为其量身打造了一个颇具颠覆意义的摩登新形象。入口处，一袭深紫色的帷幕遮挡住视线，结合餐厅的印象派主题，给人一种神秘而现代的感觉，令人更想一探究竟。

拉开帷幕，进门首先是一个偏暗调的浪漫情怀的鸡尾酒酒吧，几何的图案、硬朗的线条、艳丽的色彩、豪华的材质、可翻牌的酒牌，以及顶面的五彩呼吸灯，都使得这里既充满娱乐氛围，又有具有别样的特色，创造出别具一格的酒吧情调。

L'Arome Fangfang is a new pub food brand that promotes bistronomy. The latest one is located on the third floor of the southern section of New Landmark Suzhou. It is a gathering place for many fashionable young people and concentrated display area for concept of urban pioneers. Nayun Design took over the store design and tailored a modern and new image that was subversive.

At the entrance, a dark purple curtain obscures the line of sight, combined with the restaurant's Impressionist theme, giving a mysterious and modern feel, making passers-by more desire to find out. Opening the curtain, the door is first of all a romantic cocktail bar with a dark tone, geometric patterns, tough lines, bright colors, luxurious materials and flopped liquor labels, as well as the top colorful breathing lights. It is full of entertainment and unique features, creating a unique bar atmosphere.

L'ARÔME
DE MA VIE

从某种意义上来说，酒吧文化其实是一个城市中产阶级的文化聚所，设计师在这个项目中希望营造出类似Lido歌舞厅的绝对质感和华丽氛围，吸取L' Atelier de Joel Rubuchon餐厅的精髓，将高饱和度色彩和装饰艺术的理念结合，半圆形沙发围合着小圆桌，主次分明而富于变化，灯影光效旋转流动之时，整个空间仿佛一个巨大的悬着幻彩的琉璃灯，流光溢彩，变幻瑰丽。

穿过酒吧，绕过人群，再拉开一道幕帘，瞬间就进入到另一个宽阔天地。这里是整个餐厅的后方就餐区，和前面幽暗的酒吧截然不同，这里音乐轻柔，光线明亮，半包沙发座椅大方舒适，整体布局让人如同置身各种美食的环抱中，有着法式餐厅典型的精致与优雅。与众不同的是，设计师通过极简的装置模式和夸张的表现手法以及具有未来感的造型设计，展现了一种太空舱既视感的用餐环境，大量的曲线运用十分具有律动感，主题色调以白色和接近白色的灰色为主，间或以红艳的玫瑰点缀，完成了抽象与幻想的有机结合，给人以新奇、科幻、有趣、不同以往的丰富体验。

In a sense, the bar culture is actually a cultural gathering place for the middle class of the city. In this project, the designer hopes to create an absolute texture and gorgeous atmosphere similar to the Lido dance hall, drawing on the essence of L'Atelier de Joel Rubuchon. Combining the concept of high saturation color with the concept of art deco, a small round table is surrounded by the semi-circular sofa. The primary and secondary colors are distinct and varied. When the light and shadow effect rotates, the whole space is like a huge hanging glass, flowing light and color, changing magnificently.

Pass through the bar, bypass the crowd, and then open a curtain, and instantly enter another wide world. This is the rear dining area of the whole restaurant. It is very different from the dark bar in front. The music is soft and bright, and the half-seat sofa seat is generous and comfortable. The overall layout is like being surrounded by various foods. It is refined and elegant as a typical French restaurant . What's different is that the designer shows a space-like dining environment with a minimalist device model, exaggerated expressions and futuristic styling. A lot of curves used are very rhythmic and the theme color is mainly white and close to white gray, and it is decorated with red roses. It completes the organic combination of abstraction and fantasy, giving people a novel, science-fiction, interesting and different rich experience.

In's 咖啡
In's Cafe

项目地点：中国台湾 / Location：Taiwan, China
项目面积：76 平方米 / Area：76 m²
设 计 师：翁新婷 / Designer：Weng Xinting

IAI 设计优胜奖

翁新婷 Weng Xinting

理丝室内设计创始人兼设计总监，视觉传达艺术与建筑的学历背景，养成对于色彩与材质细节的极高敏锐度，善于解构流行趋势与多样元素，借材料与软装的运用，发展细腻的人际与空间互动关系，诠释机能与美学并具的空间。

Lisi Interior Design Founder and Design Director
Visually convey the academic background of art and architecture, develop a high degree of sensitivity to color and material details, be good at deconstructing fashion trends and diverse elements, and develop delicate interpersonal and spatial interactions through the use of materials and soft clothing, interpreting functions and The space of aesthetics.

依山傍水、邻近台中大坑风景区，隐身在山林绿意的 In's Cafe，发想于巴黎左岸边让文人墨客至此驻足、流连忘返的咖啡厅"Cafe Voltaire"。以沉稳的普鲁士蓝与混色仿旧木纹地砖作为主体，结合欧洲常见户外咖啡厅的概念，天花板的天窗与侧墙的方格玻璃窗联结入口的玻璃门面，形成三面采光通透的玻璃屋，将室内氛围延伸至四周街景的枝繁叶茂，也将山清水秀引景入室。松木围篱墙上垂直绿意的植生墙以及洗手间前的慕斯画框装置，自然气息更显空间的诗情文艺。新艺术风格的咖啡桌椅与深橡木吧台在风化后保留做旧的余韵、营造怀旧的欧风情调，与锡板共同历经岁月的斑驳调性，清透的雕花玻璃吊灯在空间中对比出欧式样貌下的复古底蕴。

This hidden garden nearby hillsides is from a renovation of a residential garage, to construct a coffee shop encompassing greenhouse courtyard seat. The building in real palette comprises facade with moulding posts sophisticatedly designed in recessed shafts and fluted details. At the front courtyard area, the seat affords a panoramic view through colonial window grid; on the other face, the green wall attaches to pine fence panels, permeating garden breeze to the greenhouse. Entering into indoor seat,the antique solid wood kitchen attracts attention, alongside distressed tin tiles which give another complementary distinction, imbuing with a wave of nostalgia for classics.

莫迪咖啡
Cafe Modi

项目地点：中国广州/ Location : Guangzhou, China
项目面积：76 平方米 /Area : 76 m²
设 计 师：杨跃文 / Designer : Yang Yuewen

IAI 设计优胜奖

杨跃文 Yang Yuewen

广州美术学院设计学硕士。艺术与技术，亦即想象与现实本体之间的高度契合是设计师在日常事务中践行的首要原则。设计师尝试以当代艺术的视角，结合严谨的工作方式去践行某种空间想象，最终赋予项目以舒适感、艺术性及永恒性。个人曾获得意大利 A-DESIGN 设计奖、IAI 设计奖、APDC 亚太室内设计奖、艾特奖等国内外设计大奖。

Yang Yuewen, interior designer, Master of Design, Guangzhou Academy of Fine Arts.
Art and technology, that is, the high degree of fit between imagination and reality, is the first principle that designers practice in their daily affairs. Designers try to practice a certain space imagination from the perspective of contemporary art, combined with rigorous working methods, and ultimately give the project a sense of comfort, artistry and timelessness.
The individual has won the Italian A-DESIGN Design Award, IAI Design Award, APDC Asia Pacific Interior Design Award, Aite Award and other domestic and international design awards.

“我要的是短暂却完整的生命。” 出自意大利艺术家莫迪里阿尼生前说过的一句话。出生于商人家庭、生性叛逆并试图与主流意识形态划清界限的他，用他有限的生命创造了精致且独特的艺术世界。CAFE MODI 其中的 “MODI” 来自艺术家名字缩写，空间的设计线索来自艺术家相关品质：前瞻性、艺术性和当代性。

空间气质来自空间形态的介入，拱门作为相邻艺术空间的出口，设计师转劣势为优势，用大尺度的拱门造型给这个出口注入力量感，以强烈的视觉符号来赋予空间以辨识度。

“What I want is a short but complete life.” A sentence from the Italian artist Amedeo Modigliani said before. Born in a merchant family, he is rebellious and tries to draw a line from mainstream ideology, creating a refined and unique art world with his limited life. CAFE MODI's "MODI" comes from the artist's initials, and the spatial design clues come from the artist's relevant qualities: forward-looking, artistic and contemporary.
Space temperament comes from the intervention of space form, the arch as the exit of the adjacent art space, the designer turns against the trend as the advantage. The large-scale arch shape casts a sense of power to the exit, and the space is given a strong visual symbol to give the space identification. .

大乐之野谷舍精品民宿
Lost Villa Valley Land Boutique Hotel

项目地点：中国浙江 / Location :Zhejiang, China
项目面积：1500 平方米 / Area : 1500 m²
公司名称：Das Lab 设计工作室 / Organization Name : DAS LAB Design

IAI 设计优胜奖

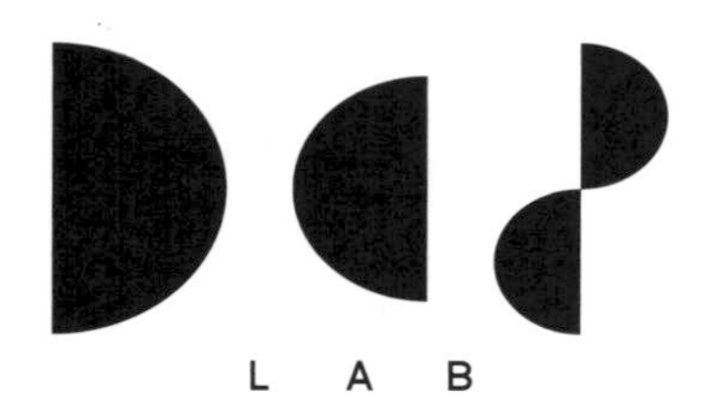

DAS Lab

DAS Lab 是一家设立于上海的多元化空间设计机构。我们用实验性的语言，在空间、品牌、产品等多个领域中通过设计发声。
DAS Lab 擅长于脱离刻板印象，习惯以别于常态的方式进行研究。通过研究，使我们更挑剔地赋予作品独特的记忆点。
DAS Lab 致力于设计概念到实体呈现，积极把控项目执行中即将呈现的细节、材质、灯光、形态等使其相互协调。高完成度的设计作品是我们始终坚持的唯一标准。

DAS Lab is a diversified spatial design studio in Shanghai. We use tentative language to speak in terms of space,product and brand.
What DAS Lab good at is giving design work distinctive and imprssive memory points by using special research methods.
Also, DAS Lab is devoted to high quality and coordinated design from concept to real according to details like form, material, lighting and so on.

餐厅是本项目最具复杂性的空间。该空间包含了聚会、用餐、入住办理、洽谈等功能。由三个高低不一的建筑组合而成，中段留有10米的挑高空间，两边是8米落地窗。面对繁茂的植被和图画感极强的自然景观，我们希望通过材料的物理性营造人工与自然的高强度碰撞以及一种直觉的、潜意识的、肌体上的平衡感。并在内部创造出与外部一样的开放性，提供一瞬间便令人印象深刻的休憩之所。

Dining room is the most complex space in this project, including functions as throwing parties, having meals, check-in and business meetings. It is made up of three various buildings and the middle section has a height of 10 meters and both sides have French windows of 8 meters high. In the face of luxuriant vegetation and very graphic natural landscapes, we hope to create high-intensity clashes of artificial work and nature, and a balance of intuition, subconscious and body through the physical properties of materials. We also want to internally create openness same as exterior and provide an impressive rest environment in an instant.

关中原宿
Guanzhong Yuansuo

项目地点：中国陕西 / Location :Shanxi, China
项目面积：4000 平方米 /Area : 4000 m²
设 计 师：李硕 / Designer : Li Shuo

IAI 设计优胜奖

李硕 Li Shuo

北京水石空间环境设计有限公司联合创始人
S+STTJ 设计工作室创意设计师
从事室内设计、展览展示设计、建筑设计及园林设计、品牌策划及服装设计。
曾主持设计过别墅、会所、商业空间、酒店、建筑、展览展示、庭院及大型厂区规划等。
经过 18 年的时间历练，也随着经验和眼界的增长与积累，从艺术与社会的角度对设计有了更深层次的、也更根本的理解。

Beijing's Shui and Shi Space Environment Design Co. Ltd. co-founder
S+STTJ design studio creative designer
Engaged in interior design, exhibition design, building facade design and landscape design.Has presided over the design of villas, clubs, commercial space, hotel, building facade renovation, exhibition design, garden and large-scale factory planning, etc..
After 18 years of experience, with the growth and accumulation of experience and vision,From the perspective of art and society, it has a deeper and more fundamental understanding.

此民宿坐落于陕西咸阳的袁家村，自古称陕西的中部为“关中”，而酒店的建筑尽量保留原有建筑的外观及结构，使之成为古朴的庭院民宿，故为“关中原宿”。

整个民宿设有豪华套房、家庭房、套房。以及陕西特有炕房等共11种房型，以便满足客人的多种选择。在有限的空间内，还设置了会议室、KTV、健身房、餐厅、酒吧等娱乐设施。

This lodge is located in Yuan Jia Village in Xianyang, Shaanxi. The central part of Shaanxi is called "Guanzhong" from ancient times. The hotel's buildings try to preserve the appearance and structure of the original buildings as far as they are.
There are 11 types of apartments, such as deluxe suites, family rooms, suites, and Shaanxi special Kang rooms, in order to meet the guests' choices. In the limited space, there are also meeting rooms, KTV, gym, restaurant, bar and other entertainment facilities.

L 酒店
L Hotel

项目地点：中国重庆 / Location :Chongqing, China
项目面积：900 平方米 /Area : 900 m²
设 计 师：尹晨光 / Designer : Yin Chenguang

IAI 设计优胜奖

尹晨光 Yin Chenguang

重庆非间室内设计有限公司（VSD DESIGN）创始人及设计总监，叁仟界艺术设计研究院（重庆）有限公司设计总监，APDF 亚太设计师联盟专业会员。

简单、真实、专注面对人、物、事，设计是一门修心的艺术，让您不断发现问题解决问题。

VSD DESIGN Founder and design director Design Director, San Qian Jie Art Design and Research Institute (Chongqing) Co., Ltd.
APDF Asia Pacific Designers Alliance Professional Member.

Simple, true, and focused on people, objects, things . Design is a masterpiece of art that allows you to constantly discover problems and solve problems

RMK酒吧
RMK Bar

项目地点：中国深圳 / Location :Shenzhen, China
项目面积：74 平方米 / Area : 74 m²
设 计 师：王琨 / Designer : Kevin Wang

IAI 设计优胜奖

王锟 Kevin Wang

国内知名空间设计大师，艺鼎设计创始人兼首席设计师。坚持完美有趣的设计。
凭借敏锐的商业视角和深厚的美学素养，引导客户创造全新的消费体验，作品多次荣获新加坡 INSIDE 年度最佳餐饮空间、英国 WINA 年度最佳餐饮空间、艾特奖等行业权威奖项的认可。

A well-known domestic space design master, founder and chief designer of Yiding Design, insists on perfect and interesting design. With a keen business perspective and profound aesthetics, it guides customers to create a new consumer experience. The work has won many awards such as Singapore INSIDE's annual best dining space, UK WINA's annual best dining space, and Aite Award.

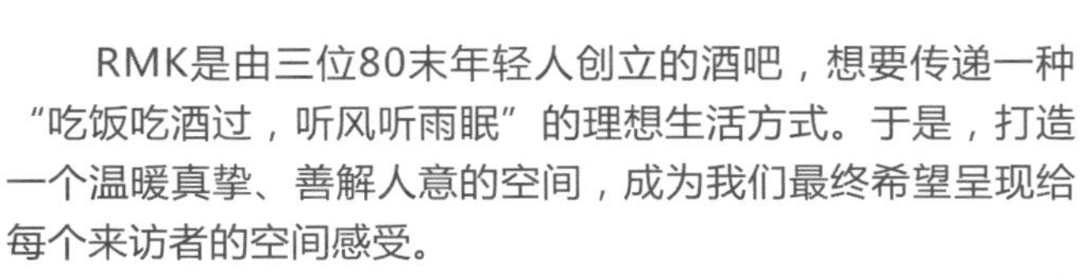

RMK是由三位80末年轻人创立的酒吧，想要传递一种“吃饭吃酒过，听风听雨眠”的理想生活方式。于是，打造一个温暖真挚、善解人意的空间，成为我们最终希望呈现给每个来访者的空间感受。

艺鼎以日本武士为空间概念，以极具质感的原始色彩为空间基调，营造轻松、自然而又富有个性的空间氛围。如何在6米层高、74平方米的狭小空间，扩大空间在视觉上的尺度感，成了考验设计师的一大挑战。艺鼎尝试使用“打破”和“重组”的方法，让空间构造尽量简单而直接，在限定的范围内营造出让受众过目难忘的视觉效果。

Founded by three post-1980s young men, RMK wants to deliver an ideal lifestyle featuring “quietly sleeping whilst listening to the wind and rain after eating and drinking” . Therefore, creating a warm, cozy and considerate space is what we ultimately hope to present to each visitor.

With the Japanese samurai as the concept of space, Yiding Design uses the original color featuring the strong sense of texture as the keynote to create a relaxed, natural and individual space. How to expand the spatial sense of scale in such a narrow space (6m in height and 74 m2 in area) has become a challenge for designers. However, Yiding Design tried to use the method of “breaking” and “reorganizing” to make the space structure as simple and direct as possible, and to create a visual effect unforgettable to visitors within a limited space.

富力希雨顿度假酒店
Hilton Huizhou Longmen Resort

项目地点：中国广东 / Location : Guangdong, China
项目面积：45000 平方米 /Area : 45000 m²
设 计 师：周涛 / Designer : Zhou Tao

IAI 设计优胜奖

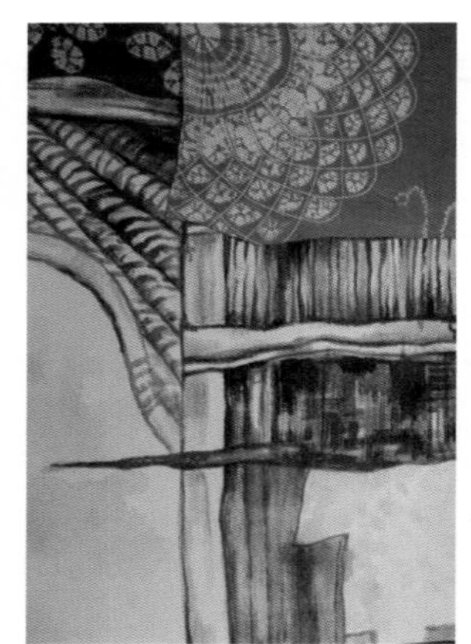

周涛 Zhou Tao

知名酒店设计师，全球酒店设计界最高荣誉 Gold Key Awards（金钥匙奖）设计大奖获得者。带领团队创建利昂设计（简称 LEOGD），并于 2015 年成功挂牌上市成为全球华人酒店设计第一股，凭借深厚的专业知识及丰富的项目经验当之无愧地成为亚洲领军性酒店行业意见领导者。主笔设计的惠州龙门富力希尔顿度假酒店等项目连续获得多项国际重量级大奖。

The well-know hotel designer. Gold Key Awards winner. Found LEOGD and successfully listed in 2015 to become the first Chinese hotel design stock in the world. With profound expertise and rich project experience he has become the leading opinion leader of Asia's hotel industry. Projects designed by his lead, such as HILTON HUIZHOU LONGMEN RESORT, have won a number of international heavy-weight awards and honors.

砖，也从来不是一种简单的材料，它古老而神奇，老砖中甚至还遗留着数十年前的人在童年儿时的随手涂鸦，在这个空间里找到了过去的记忆和承载记忆的形式。设计师将旧址房屋遗留的六万多块百年老砖全部用在了酒店的墙面建筑，不仅在许多区域呈现出了独特的视觉效果，更从成本上直接为业主投资方节省了近三千多万的材料建造成本。

竹乃中华文化之高洁者，酒店所在区拥有大面积的竹林，设计师选择就地取材，因地制宜，采用山上环保的竹子，通过经验丰富的当地工匠对竹子经过复杂而精湛的特殊技术处理装饰成竹墙、竹灯等，休闲静逸的休闲区与户外远山的繁茂绿林相呼应，在阵阵微风下竹灯轻轻摇摆，让人仿佛身临青山竹林之中，享受天然氧吧。

设计师还请来几十位当地的农民艺术家，根据提供的设计概念，亲手绘画出代表着最本色而古老的当地文化特色的装饰画，用于酒店软装的各个区域，处处体现客家文化情怀的印象，以及自然与人文之间有情感响应的协调互融的关系，形成一种独特的审美观念和文化传承，达到质、形、色的完美统一。每一个文化的融入，都是对人类历史变迁的印迹保护，设计解构生态文明的再生行为意义。

Brick is never as simple as it looks. With its ancient and miraculous myth, the antique bricks even were marked with scrawl and handwriting in someone' s childhood. A time tunnel through the past is transformed in a special way via all these antique bricks. Sourced locally, designers were drawn on bamboos and its growth morphology in mountain. By inviting experienced local craftsman to further study, these bamboos are sophisticatedly and exquisitely ornate as bamboo wall, wall sconce and pendent lights. Guests see beyond outdoor view in this tranquil leisure lobby tea lounge, bamboo lights overhead slightly swing in breeze, which seem to navigate guests into the nature and a fresh air bar. In purpose of fully delivered its initial design concept on Hakka culture, dozens of local farmer artists were invited to paint that represented the most original and antiquated Longmen elements. Furniture, fixture and equipment used in different zones in this hotel are tucked away all feeling on Hakka culture, intimate connection and cohesion between nature and humanity. Unique aesthetic perception is rooted here with profound reunion on its nature, formality and performance. Every inherited culture detail in design is carved with human being' s subtle protection on history. In design, decoding of ecological civilization is the meaning of regeneration and innovation.

后海薇酒店
Houhai Vue Hotel

项目地点：中国北京 / Location :Beijing, China
项目面积：10000 平方米 /Area : 10000 m²
公司名称 / Organization Name : Ministry of Design Pte Ltd

IAI 设计优胜奖

MINISTRY OF DESIGN

Ministry of Design

Ministry of Design 是一家综合的设计公司，可以提供建筑、室内和品牌设计，曾两次获得新加坡总统奖，三次获得纽约金钥匙奖，并被国际设计奖（美国）评为"年度设计师"，并经常在《Wallpaper》《Frame》《Surface》的杂志上面发表作品。

Ministry of Design is a comprehensive design company that offers architectural, interior and brand design. It has twice won the Singapore Presidential Award, three times the New York Golden Key Award, and was named "Designer of the Year" by the International Design Awards (USA). And often published in the magazines of Wallpaper, Frame, Surface.

"薇"酒店的旗舰店位于北京后海的胡同区。它坐落在风景如画的后海湖边，毗邻后海公园以及历史悠久的老北京胡同，胡同里一直有当地居民居住并保持着当年的风貌。酒店不远处则是著名的北京后海酒吧街。

作为一个改造项目，酒店整个园区是由一系列19世纪50年代的历史建筑组成，设计师对这些建筑进行艺术处理和改造，成就了具有"多面"的园区环境。客人可以从酒店园区内发掘不同的空间，包括一系列的园林景观、一个面向繁忙的胡同路的面包咖啡房、一个可以俯瞰后海的屋顶酒吧、设施齐全的健身房以及八十多间客房和套房，其中一些房间还配有独立花园或者享有俯瞰后海的视野。

The flagship store of Wei Hotel is located in the Hutong District of Houhai, Beijing. It is situated on the shore of the picturesque Houhai Lake, adjacent to Houhai Park and the old Beijing Hutong with a long history. The Hutong has been inhabited by local residents and maintains its original style. Not far from the hotel is the famous Houhai Bar Street in Beijing.

As a renovation project, the whole hotel park is composed of a series of historical buildings in the 1950s. Designers have made artistic treatment and renovation of these buildings, which has created a "multi-faceted" Park environment. Guests can explore different spaces from the hotel park, including a series of landscape, a bakery cafe facing busy alleys, a roof bar overlooking Houhai, a well-equipped gymnasium and more than 80 rooms and suites, some of which are equipped with independent gardens or enjoy a view overlooking Houhai.

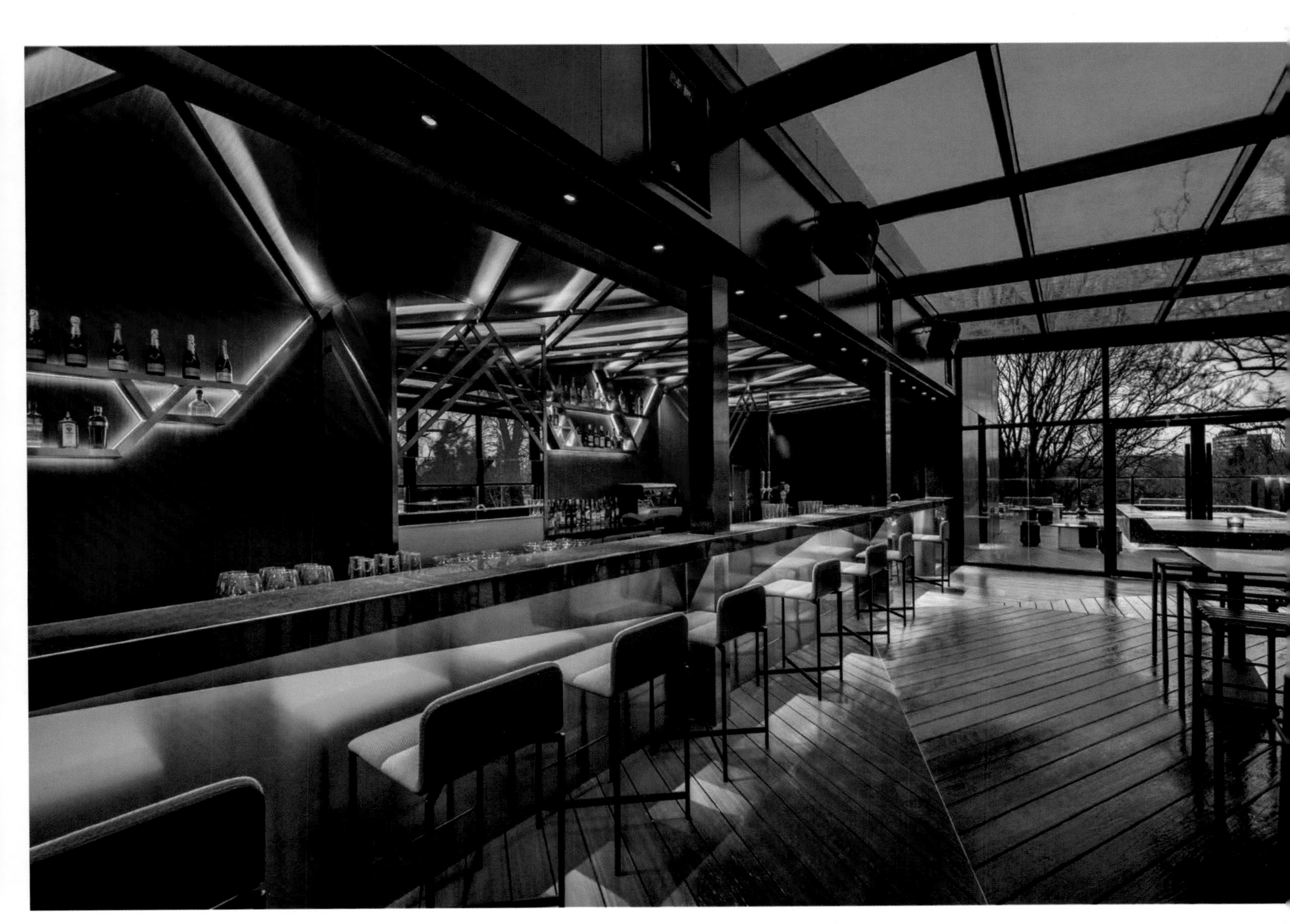

Entertentement and Clubs Space

室内 —— 娱乐/会所空间

X.俱乐部
X.Club

项目地点：中国广州 / Location : Guangzhou, China
项目面积：1166平方米 / Area : 1166 m²
设 计 师：黄永才 /Designer : Ray Wong

IAI 最佳设计大奖

黄永才 Ray Wong

RED设计创办人、设计总监
共和都市合伙人、设计总监
作品曾获多想国际国内奖项

Ray Evolution Design;
Founder & Design Director
Gonghe Dushi Partners& Design Diector;
Works has won many domestic and international awards

“X”是未知，充满各种可能性。

“X”是拒绝，拒绝没有想象的无趣。

远离地面很有趣。X.俱乐部希望富有想象力、情绪化和需求驱动，例如尝试创建云漫步和体验。

从入口进来便可见到有白色的“GRG”云团充斥着一侧的空间。云团由近到远、从高至低延续至舞台前端。由高至低的“云团”的设计使得所有云团里的客人都能看见及感受到舞台的热度。深黑色的天花与地面如浩瀚的宇宙，使得白色的云团清晰灵动。每处云团的上空设计了无序的LED点灯，营造繁星点点的空。

“X” indicates unprediction,which is full of possibilities.
“X” indicates refusal,which refuses any unimaginative vapidity.
It's interesting to stay away from the ground. X - club wants to be imaginative, emotional and demand - driven, such as trying to create a cloud walk and experience.
X-CLUB is backgrounded by walking-in-air reverie to trace out the poetry-like scene for the upcoming. Floating cloud in the Van Gogh' s starry sky and pleaceful people and so on.
While entering X-CLUB," GRG" white cloud emerges and full-fill one-side space. The cloud,from far to near and from top to bottom,extends to the front of the stage.High-to-low cloud cluster design makes it possible for clients to get views and feel the heat from the stage. Backgrounded by dark black cosmic-like ceiling and floor, the clouds look more clear and alive.Every cloud cluster designed with decorated-light overhead is to establish a starry-sky scene.

核心筒一侧是一整面的 LED 屏幕，随着 LED 播放的主题内容的变化，"云团"染色不断地变化。光的体验丰富而梦幻。在核心筒的转角，云团在此处形成吧台，吧台上空设置了抢眼的灯光装置，比星空更加璀璨。无序却有流动感，方向指向不同的两端，一处是动感十足的舞台，一处是相当安静的静吧区。

静吧区中间由一个巨大的亚克力吧台构成，在吧台上方有密集的亚克力细管构成半通透的面体。通过投影可以看到舞台区的直播，使得静区与动区有一定的关联。在星空的天花板周围有大小不同形状各异的 LED 屏幕。播放的主题也同时影响着空间的光与色。云端漫步的遐想的概念贯穿于整个空间，使得空间有鲜明的主题及体验感。

One side of the center-pipe is all-covered by a LED screen. And dizzy-color on cloud cluster constantly change following what the subject LED screen plays. People will get an averiety and dream-like experience from the light of the clouds. At the corner of the center-pipe, a bar table gets in shape by the extended cloud. Overhead the bar table, there is the catching-eye light device, while it is on work, looking more shinning than the starry sky. The device shoot out to both ends freely and smoothly, to one is in the real-dynamic stage, to the other Is in the quite bar-area.
The middle part of quiet-bar area is constracted of acrylic. Overhead it is the semi-transparent surface body made of a range of acrylic tubes. From this surface body ,after putting the view through projector, people can enjoy the live-show of the stage. So that it works to connect both the dynamic area and the quiet one,to a certain extent. Surround the sky-ceiling, there are LED screens in many sizes and shapes. The subjects playing on the screens also effects on scene of the whole interior space. The concept of walking-in-air reverie does throughout the whole space. While people walks in X-CLUB, they can feel the vivid theme and definite experience.

中南樾府
Zhongnan Mansion Club House

项目地点：中国杭州 / Location :Hangzhou, China
项目面积：3050 平方米 / Area : 3050 m²
设 计 师：凌子达 / Designer : Kris Lin

IAI 设计优胜奖

凌子达 Kris Lin

取得法国 CNAM 建筑管理硕士学位。2001 年在上海成立了"KLID 达观国际设计事务所"。2006 出版个人作品集《达观视界》。共累计荣获全球奖项 692 项；在 A'design award 获奖设计师排行榜、世界设计排行榜中，凭借 210 分的好成绩，获得全球第三名，华人中国第一名殊荣由 International Design Awards 的 Winner Ranking 统计，凭借 1400 分的总分，在全球获奖设计师中获得"全球第八名"、室内建筑部分"中国第一名"的殊荣。

Received a master's degree in building management from CNAM, France. In 2001, "KLID Daguan International Design Office" was established in Shanghai.
2006 Published a collection of personal works "Da Guan Vision".
A total of 692 global awards have been won;
In the A'design award award-winning designer rankings and world design rankings, with a score of 210 points, the third place in the world, the Chinese first place in China by the International Design Awards Winner Ranking statistics, with a total of 1400 points Points, won the [World's eighth place], indoor building part [China's first place] in the world's award-winning designers.

该项目是一个隶属住宅区内的会所，提供住宅区会所的功能，主要的活动空间都在B1层，所有功能覆盖了：咖啡厅、健身房、瑜伽教室、图书馆、茶会所、儿童学习区、地下庭院等不同的会所空间。目标是能满足老、中、青三代人的需求。

在室内设计和地下室的景观，及建筑设计中，对方提供了很大的空间，需要对整体进行一个设计构思和突出功能区的特色。

The project is a residential area under the center, to provide the function of the residential clubhouse, the main activity space on the B1 floor, covers all functions: coffee shop, gym, yoga room, library, tea party and children study area by different functional areas, including different clubs such as the underground yard space. The goal is to meet the needs of three generations.
Interior design and basement landscape, and architectural design. The other side provides a large space, which requires a design conception of the whole and highlights the features of the functional area.

城市中的绿节点
A Green Node

项目地点：中国台湾 / Location :Taiwan, China
项目面积：330 平方米 /Area : 330 m²
设 计 师：陈鹏旭 / Designer : Steve Chen

IAI 设计优胜奖

陈鹏旭 Steve Chen

空间美学室内装修有限公司，提倡“地球永续+空间美学+人本健康”，重视人与自然间的平衡，并确保空间使用者健康。陈鹏旭设计总监以“空间美学，人本健康”为公司的设计理念，实现健康的空间美学设计。

Design Director Peng Xu Chen of Aesthetic of Space Design Co., Ltd. advocates "Earth Sustainability + Space Aesthetics + Human Health", focusing on not only balance between human and nature, but also the health of space users. He insists "Space Aesthetics with Human Health" as company's philosophy to achieve healthy spatial aesthetics design.

这是一幢两层楼高的建筑，两翼都有有限的窗户，可以阻挡西部露天和繁忙的街道。中央轴线由大窗户系统通过楼梯和窗户末端与露天庭院融合。在这个庭院里面，我们特意种了一棵巨大的树，创造了自然的焦点和场景。高架地板设计将停车位置于一楼并连接到庭院，因此烟囱效应可以对车辆进行排热。带有大型玻璃面板窗户的主轴带来自然光线，营造出明亮的陈列室。

走上楼梯，从高高的天花板玻璃顶部用绳子吊住的下垂的空气植物（Tillandsia）不仅隐隐有一些绿色，还可以看作垂直雕塑，一个软化空间的枝形吊灯。此外，在楼梯下面种植了一个覆盖地面的草坪地毯，从草坪外面到庭院中的树木，形成了一条连续的绿色走廊。

It' s a two story tall building with limited window opening on the both wing to block the western exposure and busy street. The central axis defined by large window system at front through staircase and end with an open air court yard. Inside this courtyard, we specially planted a huge tree to create the focal point and scene of nature. The raised floor design to put parking at first floor and connect to the court yard thus chimney effect can carry out heat from vehicles. Main axis with large glass panel windows bring in natural lights to create bright showroom.

Step on the stairs, air plants (Tillandsia) hanging down from high ceiling glass top hold by strings not only give some hint of green but can also be seen as vertical sculpture, a chandelier that soften the space. Furthermore a mall groundcover carpet is planted underneath the staircase creates a continuous green corridor from outside lawn to the tree in the courtyard.

墨花香
Boundless-Where Humanities Meet Community

项目地点：中国台湾 / Location :Taiwan, China
项目面积：680 平方米 /Area : 680 m^2
设 计 师：陈鹏旭 / Designer : Chen Pengxu

IAI 设计优胜奖

该选址位于台北市著名的哲学社区，毗邻台湾师范大学（NTNU）。这个区域不仅是许多学者的家，也是独立书店的集群，画廊为这个地区提供了特别的东西。受周围环境的启发，我们试图模糊物理边界，并将文化和诗意的氛围引入公共空间。从外面的拱廊到正门进入接待处，有一些大木板模仿部分书籍。高高的天花板大堂周围有书架，可作为图书馆，让居住者和游客可以阅读或召开社区读书俱乐部活动。书架形成了一个半透明的分隔物，把大厅分开成半私人电梯厅和邮件室。

烧烤图案和镜面反射是实现设计师实用目的的理念的另外两个关键要素。地下停车场、二楼健身、瑜伽室和会议室的木质涂装金属格栅天花板使这些公共空间符合消防安全规范。洒水系统和烟雾探测器可以隐藏在格栅内。但这些也带来了大量书籍的印象效果。

The site is a mixed use apartment building situated in the famous philosophical neighborhood in Taipei city adjacent to the National Taiwan Normal University (NTNU) where school teachers are educated. This area is not only home for many scholars, but also clusters for independent bookstores, galleries give special something to this area.

Inspired by its surroundings, we try to blurred the physical boundary and introduce cultural and poetic atmosphere into common spaces. From the outside arcade to main entrance into reception, there are sections of big wooden panels imitate section of books. The high ceiling lobby is surrounded by book shelves can serve as a library allow occupants and visitors to pick up something to read or even host community book club events. A translucent divider creates by bookshelves separate lobby to semiprivate elevator hall and mail room.

Grill pattern and mirror refection are the other two key elements to carry the out the designer' s concept with practical purpose. The metal grill ceiling with wood finish at basement parking lot, the second floor fetidness/yoga room, and meeting room allow these common spaces to fit the fire safety code. Sprinkle system and smoke detector can hide within grills. But these also bring in the impression of pilled books.

恒茂未来都会
Future City Sales Center

项目地点：中国江西 / Location :Jiangxi, China
项目面积：5600 平方米 / Area : 5600 m²
设 计 师：刘杰 / Designer : Jemes Liu

IAI 设计优胜奖

刘杰 James Liu

AOD 艾地设计创始人、设计总监
亚太建筑师与室内设计师联盟理事

十余年来致力于中国的高端室内设计与建筑的集成，以其深厚的艺术修养和丰富的设计经验，创作出极具国际影响力的设计作品。作品荣获意大利 A' design award 金奖、伦敦 SBID 国际设计大奖、美国 IDA 国际设计大奖、法国双面神、IAI 全球设计奖等国际和亚太区设计奖项。

AOD Ai Di Design Founder, Design Director
Asia Pacific Association of Architects and Interior Designers
For more than ten years, he has been dedicated to the integration of high-end interior design and architecture in China. With his profound artistic accomplishment and rich design experience, he has created internationally influential design works. His works have won international and Asia-Pacific design awards such as the Italian A'design award gold medal, London SBID International Design Award, American IDA International Design Award, French Double-faced God, IAI Design Award.

设计随着人类需求的变化而创新，在新中式风格的售楼处不断涌现的当下，拥有大量欧美项目开发背景的恒茂集团意欲打造一个现代、生态、科技且与众不同的销售中心，这与AOD推陈出新的设计思维相契合。本案AOD将建筑与室内集成设计的理念贯穿项目设计的始终，以创新的设计概念奠定了项目的整体设计基调；用时尚动感的弧形线条勾勒出科技感；整个空间自然流畅、充满时尚现代的气息。

Recently many sales centers with new Chinese style have continuously appeared, the design should be innovated according to constant changes of Heng Mao Group demands. With lots of European and American projects, HENGMAO GROUP wants to build a modern, ecological, technological and distinctive sales center, which corresponds to the original design concept of AOD. AOD makes the concept of building and indoor integration design throughout this project and set the overall design style: use fashionable and dynamic arc-shaped lines to show the sense of technology; the whole space is modern and fashionable that makes you comfort

四季草堂会所
Four Seasons Cottage Villa Club

项目地点：中国上海 / Location :Shanghai, China
项目面积：1200 平方米 / Area : 1200 m²
设 计 师：王建 /Designer : Wang Jian

王建 Wang Jian

从业 23 年
国家注册室内设计师
中国高级室内建筑师
中国建筑协会室内设计分会会员
中国建筑装饰协会会员
中国室内装饰协会会员
全国杰出中青年室内建筑师
全国住宅装饰装修行业优秀设计师

23 years in business
National registered interior designer
Chinese senior interior architect
Member of China Architecture Association Interior Design Branch
Member of China Building Decoration Association
Member of China Interior Decoration Association
National Outstanding Young and Middle-aged Interior Architect
Excellent designer in the national residential decoration industry

"四季草堂"顾名思义，旨在以传统文化和皇权底蕴为依托，打造一个承载中国历史文化价值的新典范空间。整体设计体现中国传统的端庄，并凸显了少见的中国古代威严和大气磅礴的一面。结合现代人的审美视角，对古典内蕴重新审视，主动寻回设计的文化自觉意识，秉承与古代中国园林相仿的设计手法。传统古典中式对于大空间的驾驭拥有上千年历史的沉淀，而现在已成为一种生活时尚、文化情调，以一种返璞归真的姿态出现在人们面前。

Just as its name implies, "The Four Seasons Cottage" will realize that the design of this case is dependant on the traditional culture and the imperial power details to create a new model space that bears the weight of the historical and cultural value of China. The overall design embraces the dignity of traditional China and highlights the rare majesty and great momentum of ancient China. Combining the aesthetic perspective of modern people, re-examining the classical intrinsic and actively retrieving the cultural consciousness of design, adhereing to the design method similar to the ancient Chinese garden, applied in the design of this luxury villa. The traditional classical Chinese style has been reining the large space for a thousand years of history deposits. however now it has become a kind of life style, a cultural sentiment, and it appears in front of people with a recovering its original simplicity posture。

Tan Y
Tan Y

项目地点：中国南京 / Location : Nanjing, China
项目面积：1000 平方米 /Area : 1000 m²
设 计 师：王伟 / Designer : Wang Wei

IAI 设计优胜奖

王伟 Wang Wei

道伟（IDDW）室内设计有限公司
高级室内建筑师、建筑摄影师
CIID 中国建筑学会室内设计分会会员
AIDIA 亚洲室内设计联合会专业代表
APDF 亚太设计师联盟会员
IFI 国际室内建筑师联盟会员
JSIID 江苏省优秀室内设计师称号
CIID 中国室内官方推荐设计师
中国地区 50 位优秀青年室内设计师
大师对话营推荐青年设计师

IDDW Interior Design Co., Ltd.
Senior Interior Architect and Architectural Photographer
CIID China Institute of Interior Design Branch Member
Professional Representative of AIDIA Asian Interior Design Federation
APDF Asia Pacific Designers Association Member
IFI International Interior Architects Alliance Member
JSIID JiangSu Province Excellent Interior Designer Title
CIID China Indoor Official Recommendation Designer
50 Excellent Young Interior Designers in China
Master Dialogue Camp recommends young designers

设计的根本是让人们去享受，而享受是内在的自我与外在的世界产生连接后的体验。设计师用低调的色彩与自然元素，将空间作为人的背景，让人们置身于此时，能够抛却浮华与形式主义，回归到交流本身，与空间互动，与他人互动，这才是人与空间最适当的平衡。

设计师在灰调的基础上搭配定制的软装，使充满现代感的舒适流淌在每个角落，细节之处的完美，更能体现设计师对于设计的至臻至诚。我们用心发现生活的真意，使新的休闲模式演变成品味生活的一种艺术。

在这里，人们摒弃压力，游弋心灵的港湾。

The essence of design is to let people enjoy, and enjoy is the experience of connecting the inner self with the external world. Designers use low-key colors and natural elements to make space a human background, allowing people to leave their glitz and formalism at this time, return to communication itself, interact with space, and interact with others. This is the most proper balance between the people and space.
The designer matching custom-made soft decoration on the basis of the gray tone makes the comfort which is full of modern feeling flowing in every corner, the perfection of the details, and even more reflects the designer's perfectest and sincerity to the design. We use our heart to discover the true meaning of life and make the new leisure model evolve into an art of life.
Here, people abandon the pressure and swim in the harbor of the soul.

金龙大酒店龙壶茶饮会所
Golden Dragon Hotel Dragon Pot Tea Club

项目地点：中国四川 / Location :Sichuan, China
项目面积：1000 平方米 / Area :1000 m²
设 计 师：严俊杰 & 冯琳涵 / Designer : Yan Junjie & Feng Linhan

IAI 设计优胜奖

严俊杰 & 冯琳涵
Yan Junjie & Feng Linhan

资深室内设计师，均毕业于西南大学。独有的 CNSA 研发经历，使之在哲学与科学的高度上树立了独特的艺术思想。

Senior interior designers, both graduated from southwest university. Unique research and development experience of CNSA makes it set up a unique artistic thought in the height of philosophy and science.

本作品的设计思想致力于对中国古代建筑装饰艺术在当代的融合、复兴与发展的研究。设计主旨在于探索中国秦代、汉代、唐代建筑装饰艺术与当代室内装饰设计艺术的多元化融合，意在丰富目前中国元素室内设计里以宋、明、清建筑装饰艺术为主流的方式。

The design idea of this work is devoted to the research on the development the fusion and revival of Chinese ancient architectural decoration art in the contemporary era. The main purpose of the design is to explore the architectural decoration art of Qin, Han and Tang Dynasties in China.The diversified integration of contemporary interior design art is intended to enrich the current interior design of Chinese elements.where the art of decoration in song, Ming and Qing Dynasties are the mainstream.

茅莱山居售楼处
The Sales Office Of Maelai Hill

项目地点：中国重庆 / Location :Chongqing, China
项目面积：1200 平方米 /Area : 1200 m²
公司名称：重庆宜庆装饰设计有限公司 / Organization Name : Yiqing Design

IAI 设计优胜奖

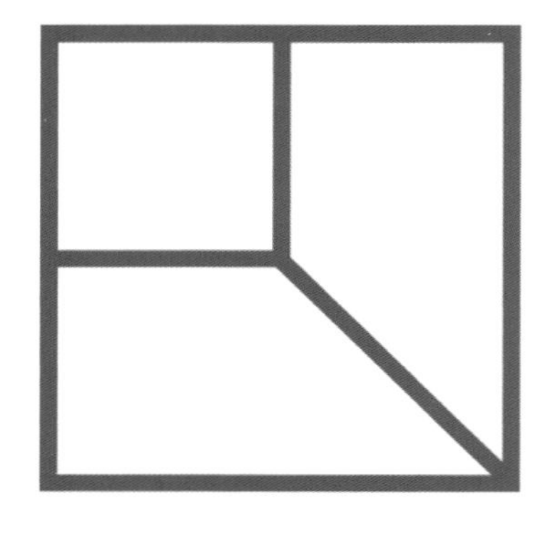

宜庆设计
YIQING DESIGN

宜庆设计 Yiqing Design

重庆宜庆装饰设计有限公司是一家服务于大型公共空间及高端住宅私人定制的专业设计公司，拥有丰富的设计及项目实施经验，并已成为多家国内大型地产开发企业的优质合作商。

Chongqing Yiqing Decoration Design Co., Ltd. is a professional design company serving large public space and high-end residential private customization, with rich experience in design and project implementation, and has become a number of domestic large real estate development enterprises of quality partners.

巴渝山水之硬朗与东方人文之雅致的碰撞。

数千年来，人类文明在适应社会发展的过程中，不断地丰富和发展着传统文化，如果摒弃根基，隔绝血脉，则将会丧失本质、迷失方向。 本案提取了巴渝文化的山水、蜀绣等传统元素符号，营造壮美意境，结合东方人文美学打造充满想象力的个性空间。细节处以松树、祥云、山影白云、水纹、山石等元素彰显“修身齐家”的优雅气息，古典杂糅着现代，与磅礴大气的整体空间交相辉映。此处可观山水，亦可捧书品茶，气定神闲。这里，不仅仅是销售中心。

“诚者天之道也，思诚者人之道也。”——《孟子·离娄上》

For thousands of years, human civilization has been constantly enriching and developing traditional culture in the process of adapting to social development. If the foundation is abandoned and blood is cut off, it will lose its essence and direction. This case has extracted the symbols of traditional elements such as landscape and shu embroidery from bayu culture, created a magnificent artistic conception, and combined with Oriental humanistic aesthetics to create an imaginative individual space. Details such as pine trees, auspicious clouds, mountain shadows and clouds, water patterns and rocks reveal the elegant atmosphere of "self-cultivation and family". Classical blending with the modern, it is reflected in the magnificent overall space.

越秀星汇文华
Yuexiu Starry Manwah

项目地点：中国广州 / Location :Guangzhou, China
项目面积：320 平方米 /Area : 320 m²
设 计 师：周涛 / Designer : Zhou Tao

IAI 设计优胜奖

周涛 Zhou Tao

知名酒店设计师，全球酒店设计界最高荣誉金钥匙奖（Gold Key Awards）设计大奖获得者。带领团队创建利昂设计（简称LEOGD），并于 2015 年成功挂牌上市成为全球华人酒店设计第一股，凭借深厚的专业知识及丰富的项目经验当之无愧的成为亚洲领军性酒店行业意见领导者。主笔设计的惠州龙门富力希尔顿度假酒店等项目连续获得多项国际重量级大奖。

The well-know hotel designer. Gold Key Awards winner. Found LEOGD and successfully listed in 2015 to become the first Chinese hotel design stock in the world. With profound expertise and rich project experience he has become the leading opinion leader of Asia's hotel industry. Projects designed by his lead, such as HILTON HUIZHOU LONGMEN RESORT, have won a number of international heavyweight awards and honors.

利昂设计团队精心打造“家宴”主题，“家”与“宴”，一个让居住者几代同堂都能舒适生活的居所，也是一个可宴请多方亲朋和商友聚会的便利空间。

the Leogd design team elaborately created the theme of “family banquet”，“home” and “banquet”，a residence that allows residents to live comfortably for generations, and is a feast for friends and family, also a convenient space for gatherings of business friends. In design, each layer of a home

G空间
G Space

项目地点：中国台湾 / Location :Taiwan, China
项目面积：396 平方米 / Area :396 m²
设 计 师：蔡明宏 / Designer :Tsai Minghong

IAI 设计优胜奖

蔡明宏 Tsai Minghong

汉玥室内设计
创始人兼设计总监、主持设计师

中国语文学院的艺文背景与木工匠心的实务薰陶，作品视野包罗国际眼界，注重使用者和空间的对话与互动，激荡现代实验派精神，藉由室内设计将诗意与美学历练带入美好生活中。

Han Yue Interior Design Founder & Design Director

With educational background of Chinese literature and artisanal experience, Tsai Ming Hong's works encompasses user's interaction with spatial flows. As a experimentalism designer, Tsai emerges new ideas and imbues lifestyle with poetic undertones.

传统的发廊经常将冲洗区设在内部的后端，以便于管线的安排。设计师讨论后认为，将冲洗区放在配置的中心位置,可以利用不同的平面配置来呈现空间的层次与力量。在这个意义之下，空间的操作与呈现会展现出平常没有的分割与区域，创造出不同的分区。

这是一个挑高8米的基地，为了在里面表现空间感，将每个分区利用分割与放置块体的概念来操作，使得空间较有层次感与乐趣性。除此之外，也利用灯光效果制作近三百颗LED组成波浪意象的灯海，透过灯海传达出头发润泽与柔顺的意念。

In traditional designs the rinse area often been putted in the back of the salon so that the pipes are easier to be arranged. In this project designer reckons that to put the rinse area in the center of the salon could make level and strength by using different plane configuration after discussion. In this case, the space will show different allocated area by the design.
In addition to this project is a 8 meters site, by using divided districts and putting cubes to make this space more multilevel and functional pleasure. Also, designer by using almost 300 LED bulbs to create the image of hair waves in the air.

Wee俱乐部
Wee Club

项目地点：中国云南 / Location :Yunnan, China
项目面积：1000 平方米 /Area : 1000 m²
设 计 师：方飞 / Designer : Fang Fei

IAI 设计优胜奖

方飞 Fang Fei

2018 红棉中国设计奖至尊奖
2018 英国 SBID 室内设计大奖
2018 IAI 全球设计大奖优胜奖
2018 IDS 国际设计先锋榜金奖提名
2017 第八届中国国际空间大赛铜奖
2015-2016 中国设计星 全国十二强
2016 CIID 第十九届空间设计大赛银奖
台湾 2016TAKAO 室内设计大赛银奖
2016 第四届中国营造空间设计大赛金奖
第四届 ID+G 金创意奖国际空间实际大赛专业类金奖
Modern Gym Style 中国健身空间设计大赛金奖
2018 Cotton Tree China Design Award Supreme Award
2018 British SBID Interior Design Award
2018 IAI Design Awards Excellence Winners
2018 IDS International Design Pioneer Gold Nomination
2017 Eighth China International Space Competition Bronze Award
2015-2016 China Design Star National Top 12
2016 CIID 19th Space Design Competition Silver Award
Taiwan 2016TAKAO Interior Design Competition Silver Award
2016 The 4th China Space Design Competition Gold Award
The 4th ID+G Gold Creative Awards International Space Reality Competition Professional Gold Award
Modern Gym Style China Fitness Space Design Competition Gold Award

设计师认为，WEE　CLUB的设计应该是一艘崭新的宇宙飞船，载着城市的男男女女，充满争议，又不惧争议，能够在酒精的驱使下振奋宇宙。让人忘记烦恼，忘记当下，WEE CLUB正是这群人的迷幻之地。

设计师通过三角曲面构成，将人与空间、空间与形态做镜像化处理，重新定义娱乐空间的同时，在墙面、顶面之间，产生一种构成关系，使得整个空间更具张力。

周边的水泥墙面相互渗透，让设计的基本界面变得既炫酷又绚丽。

整个娱乐空间分为A、B两个区域，A区以摇滚、嘻哈、地下风格为主要设计构思，在设计手法上通过声光电的处理，产生科技感

The designer believes that the design of WEE CLUB should be a brand new spaceship carrying the men and women of the city. It is controversial and not afraid of controversy. It can be used to drive the universe under the influence of alcohol. Forgetting troubles and forgetting the moment, WEE CLUB is the psychedelic place of this group of people.
The designer uses a triangular curved surface to mirror the human and space, space and form, redefining the entertainment space, and creating a relationship between the wall and the top surface, making the whole space more tension.
The whole entertainment space is divided into two areas, A and B. The main design concept is rock, HIP-HOP and underground style. In the design method, through the processing of sound and light,

Public Exhibition Space

室内 —— 公共展示空间

竞赛机器人实验室
Race Robotics Laboratory

项目地点：新加坡 / Location : Singapore
项目面积：3050 平方米 /Area : 3050 m^2
公司名称 / Organization Name : Ministry of Design Pte Ltd

IAI 最佳设计大奖

MINISTRY
OF
DESIGN•

MINISTRY OF DESIGN

Ministry of Design 是一家综合的设计公司，可以提供建筑、室内和品牌设计，曾两次获得新加坡总统奖，三次获得纽约金钥匙奖，并被国际设计奖（美国）评为"年度设计师"，并经常在《Wallpaper》《Frame》《Surface》等杂志上面发表作品。

Ministry of Design is a comprehensive design company that offers architectural, interior and brand design. It has twice won the Singapore Presidential Award, three times the New York Golden Key Award, and was named "Designer of the Year" by the International Design Awards (USA). And often published in the magazines of Wallpaper, Frame, Surface.

实验室空间要求具有灵活性，用来展示一系列不断变化的模块化机器人，并用于动手的培训和讲座。实验室需要一个连续的开放空间，同时要考虑小群体亲身实践的培训。为满足这一要求，MOD试图创造一个引人入胜、极具未来感空间体验，来体现工业实验室空间要求具有灵活性，用来展示一系列不断变化的模块化机器人，并用于动手的培训和讲座。

快要到达电梯厅，就会看到一个迎接观众的生动的实验室前厅。飞扬的白线穿过黑色的空间形成的网创造了一个从地面到天花扭曲变形的空间经验。从黑色信封的电梯大堂而来，一个定制的超大号的门开着，露出一个戏剧性的金属多面空间，创造了一个强烈的对比和引人注目的互补。

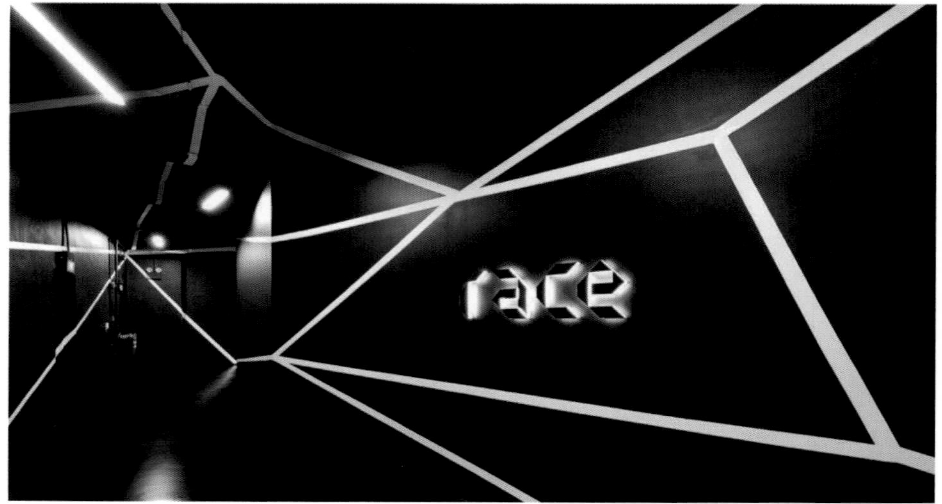

The laboratory space requires flexibility to showcase a range of ever-changing modular robots for hands-on training and lectures. The laboratory needs a continuous open space, while also considering the hands-on training of small groups. To meet this requirement, MOD seeks to create a fascinating, futuristic spatial experience that reflects the characteristics of industrial automation and precision.

As soon as you arrive at the elevator hall, you will see a lively laboratory vestibule that welcomes the audience. The flying white line through the net formed by the black space creates a space experience from the ground to the distortion of the ceiling. From the elevator lobby in the black envelope, a custom-built oversized door opens to reveal a dramatic metallic multi-faceted space that creates a strong contrast and compelling complement.

为了提供空间的最大灵活性，MOD引入了“第二层”皮肤——无缝地创造了一个动态空间——通过解构天花和墙面，形成令人炫目的多面阵列。每个面都由手工切割的空心铝管堆叠而成的模块组成，几个基本模块旋转和重复形成多个位面。这层铝管皮肤也遮盖了那些必须的机电设施，在遮挡视线的同时很容易进行操作和维护。定制发光二级管的随机布置在各个方向的面板上，体现了前沿的审美。总的来说，空间提供了极具未来感的背景来开创一个自动化和机器人技术的时代。

To provide maximum flexibility in space, MOD introduces a “second layer” of skin – seamlessly creating a dynamic space – by deconstructing ceilings and walls to create a dazzling multifaceted array. Each face is composed of a stack of hand-cut hollow aluminum tubes, and several basic modules are rotated and repeatedly formed into a plurality of planes. This aluminum tube skin also covers the necessary electromechanical facilities, making it easy to operate and maintain while obscuring the line of sight. The random arrangement of custom light-emitting diodes is on the plane of each direction, reflecting the aesthetics of the frontier. In general, space provides a futuristic background to create an era of automation and robotics.

文成堂

The Space of Wencheng Hall

项目地点：中国北京 / Location : Beijing, China
项目面积：140 平方米 /Area : 140 m²
设 计 师：欧阳昆仑、魏冰/Designer : Ouyang Kunlun , Weibing

IAI 最佳设计大奖

欧阳昆仑 Ouyang Kunlun

1994 毕业于重庆建筑工程学院
2002-2004 人民大学徐悲鸿美术学院设计艺术系空间设计专业客座教师
2003- 至今 任北京方和建筑及室内设计总监

Graduated from Chongqing Institute of Civil Engineering and Architecture in 1994

2002-2004 Department of Design Art, Xu Beihong Academy of Fine Arts, Renmin University Space Design Professional · Guest Teacher

2003-present Beijing Fanghe Building and Interior Design Director

魏冰 Wei Bing

1995 毕业于中央工艺美术学院环艺系环境设计专业学士学位
2003- 至今 任北京方和室内设计总监

Graduated from the Central Academy of Arts and Crafts in 1995 with a bachelor's degree in environmental design.
2003-present Beijing Fanghe Interior Design Director

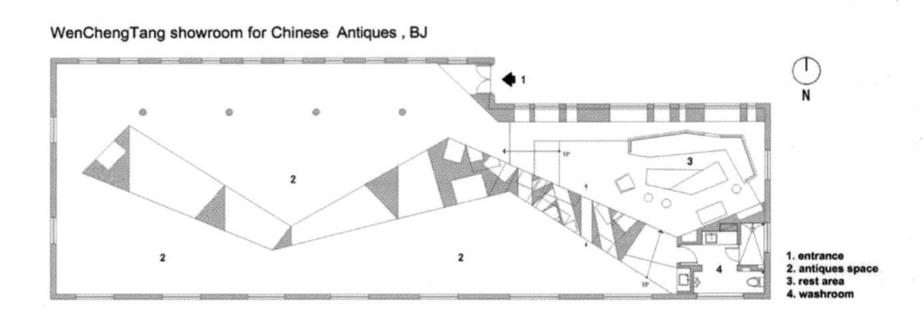

文成堂是北京最大的软木古家具行，张春林先生主持它，并渐渐追溯到古家具的源头 —— 古家具的修复和保护。

张春林先生说的最多的一句话：“没有不好的家具。”由是，他的空间里家具贯通各个时代，连接北方各个地域。

我们受邀设计展厅，仅将三件家具放入其中。总结椅凳类、桌案类、柜架类、床榻类家具尺度后，设计三个龛体，龛体高、深阔度不同，对应不同类家具，龛体于空间中心位置，连续转折如有机体，观者以游走，环绕多方位、多深度感受过去的物件。

Wencheng Hall is the largest collector of soft lumber antique furniture in Beijing. Master Zhang Chunlin is the curator and the only who gradually restored the original essence of antique furniture, their refurbishment and preservation.

In Master Zhang' s words: there is no bad furniture. As such, his collection spans different dynasties and across north China.

We' ve been invited to design the exhibit hall, and have chosen to put only 3 pieces of furniture up for exhibition. We designed 3 shrines of difference height, depth and width, corresponding to different categories of furniture: chair/stool, table/desk, cabinetry/shelf, and bed/couch. The shrines are placed in the center of the space, and are connected organically. Viewers can walk around, view from different angles and get a sense of how these furniture were used in the old days.

龛体延折至东侧，生成多向、多尺度、多组合的龛集合，众多古器物和石雕以及垂挂的烛灯，簇拥出一个交流区。三个展主家具的龛，定了空间调性，而众多小龛、密集小物件、老器物的集合就构成与三个主家具关联的场域，这里传统不孤，互相映证中国北方结构造物于形体、线条、材质、纹样……在朝代交迭中的演化。

做这个项目时，我常常想到龙门石窟。中央是卢舍那佛，左右是菩萨、罗汉、力士等等，这是一个绝美的佛国。而这个展厅，最终也汇聚几十倍于三件的家具。它们集于转折体周边，其间大小互成，构成一段古物的时空。

The three main shrines displaying the furniture, define the tone of the space, while the multitude of smaller shrines with artifacts and accessories provide the story line of how the three large pieces of furniture interact with each other. Here, the tradition of northern furniture making is manifested through shapes, lines, materials, and patterns, providing a sense of the transition from dynasty to dynasty. While working on this project, I often think of LongmenGrottoes. In the middle is Lurothe Buddha, on the left and right sides are buddhas, arhats, and heroes, forming a perfect Buddhist world. Ultimately, this exhibit hall also tries to showcase much more than 3 pieces of furniture. They gather and twist around the shrines, with contrast and harmony between the large and the small, thereby forming a space for ancient objects.

春在东方山西展厅
Spring Is The East Shanxi Exhibition Hall

项目地点：中国广州 / Location : Guangzhou, China
项目面积：1166平方米 /Area : 1166 m²
设 计 师：朱海博 / Designer : Asoka Chu

IAI 最佳设计大奖

朱海博 Asoka Chu

知名室内建筑师
深圳市室内设计师协会副秘书长
意大利米兰理工大学国际室内设计硕士
深圳市故事空间设计有限公司创始人
壹同创意集团创始人

Famous Interior Architect
Deputy secretary general of Shenzhen association of interior designs
Master of International Interior Design, Politecnico di Milano, Milan, Italy.
Founder of Shenzhen Storybox Interior Architectural Design Co., Ltd.
Founder of TO GET THE Creative Group

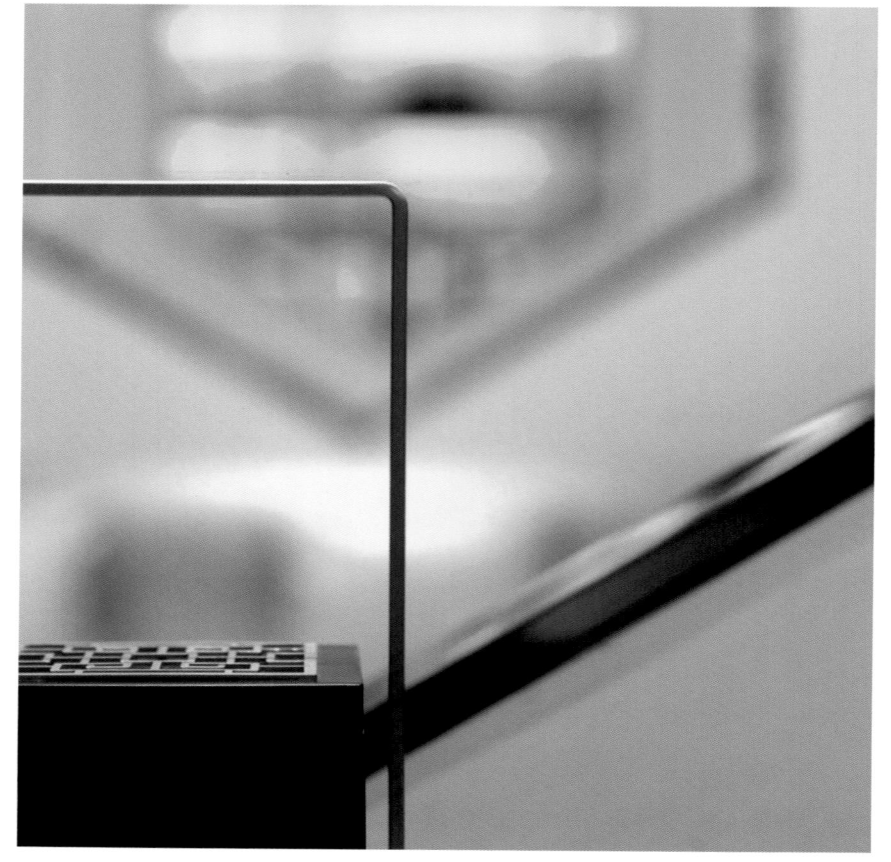

新中式，有着中国传统的韵味却比传统中式更多一份简洁和意境，更是受到年轻人的喜欢，不仅有品位，还有闹中取静的视觉哲理。

冷静，才能看清事物的本质。我们的设计调性偏向冷静与克制，这是源于我看待事物的的方式，或许也是我的生活态度。

世界缤纷多彩，所有的设计都是某种生活方式的演绎。无论是哪种生活，终将归于平淡，内里也近乎绚烂，无需故弄玄虚，无需富贵堆砌，只要魂灵的安定、纯粹和饱满。到达“不恋尘世浮华，不写红尘纷扰，不叹世风苍凉，不惹情思哀怨，闲看花开，静待花落，冷暖自知，洁净如始”的意境。空间除了满意功能性的需要，更透露出魂灵的力气，让身处其间的人悠然自得，并与之共识。中国有意境的风景融入，体现人对自然的尊重，更是东方的美、沉稳、内敛、含蓄的最佳精神气质。营造一个有温度的东方生活方式体验场景。东方人的审美：自然美学。借景、框景……

The new Chinese style, while maintaining the Chinese traditional rhythm, is supplemented with an additional portion of simplicity and artistic concept than the traditional Chinese style. It is beloved by the young people not only for its taste but also the visual philosophy of creating quietness in noisiness.
Only with calmness can we see the essence of things. Our design tonality leans toward calmness and restraints. This originates from my ways of looking at things and probably my life attitude.
As the world is colorful, each design is an interpretation of a certain lifestyle. A life of any kind will settle in insipidity. As the inner is already splendid, there is no need to mystify or have richness and nobility piled up. All needed are tranquil, pure and full souls free from attachments to the worldly matters, disturbances from the vulgarity, laments on the deteriorating public morals, additions to positive or negative affections. Such souls should also watch the blooming and falling petals with leisure and tranquility to stay self-conscious and clean as original. A space should not only offer functional satisfactions but also reveal the might of souls so that any dweller can stay leisurely and echo with the space.The blending of the artistic Chinese landscape reveals the respect for the nature, the best demeanor of the Orient featuring the beauty, calmness, restraining and implication. Create an experience scene characterized by a warm oriental lifestyle. Oriental Aesthetics: Natural Aesthetics,Borrow the scenes and frame the scenes.

东方人的审美趋势：更加开放与融合。

纵观历史，上下五千年都是一个内外融合、吸纳百川的历史。易经的易字就是变，什么都在更新迭代，唯一不变的是我们的根文化。设计师是解决问题的，但有时候问题对于甲方来说并不是问题，因为这是设计师本来就应该做的。将中国的文化引入，若隐若现，若即若离，极大地表现出意境之美。

The aesthetic trend of orientals will lean toward more openness and integration
Throughout history, the history five thousand years reflects the absorptions from assorted channels at home and abroad.
In the Book of Changes (Yi Jing), the Yi refers to changes. While everything is upgrading, the only thing that remains unchanged is our root culture.
A designer is to solve problems. However, some problems are in fact not problems for the owner of the property. For the owner, this is something you as a designer should do.
The introduction of Chinese culture is looming and lukewarm, greatly showing the beauty of artistic conception.

美的置业宁乡国宾府销售中心
The Guobin Luxury Mansion

项目地点：中国湖南 / Location : Hunan, China
项目面积：1000 平方米 /Area : 1000 m²
设 计 师：陈正茂 / Designer : Chen Zhengmao

IAI 设计优胜奖

陈正茂 Chen Zhengmao

设计执行董事、总经理，2007 年毕业于广州大学建筑设计专业，2007-2015 年 4 月就职国内大型设计公司，2015 年 5 月加盟广州达艺装饰有限公司并创立子品牌广州古德室内设计有限公司（Good Design Crew）。
有深厚的建筑及装饰设计基础，对整体项目有着完善的控制能力和出色的领导才能。

Design executive director, general manager,
Graduated from Guangzhou University in 2007 with a major in architectural design.
In April 2007-2015, he worked for a large domestic design company.
Joined Guangzhou Dayi Decoration Co., Ltd. in May 2015
And founded the sub-brand Guangzhou Good Interior Design Co., Ltd. (Good Design Crew).
With a deep foundation of architectural and decorative design, it has perfect control and excellent leadership skills for the overall project.

古德设计运用东方美学与现代设计手法，将传统装饰元素打散重组，并简化为现代的创意空间，为商业空间注入带当地文化的空间感受。

整体设计理念以山、水及人文等元素为灵感，整个空间的设计以此展开。

接待区域左右两边的背景墙，设计师运用了山峦层叠的形态，通过简化、重组形成统一而又富有层次变化的空间视觉感受，11米高的主屏风则运用了带水纹理的玉石作半透处理，石材的温润，犹如水一般让空间更润泽。

Based on Oriental aesthetics and modern design techniques, GoodDesign-Crew deconstruct and recombine traditional decorative elements, simplify them into modern creative space, and inject local culture into commercial space.
The whole design concept is inspired by the elements of mountain, water and humanity, and the design of the whole space unfolds accordingly.
The background wall designer on the left and right of the reception area used the form of mountain layer on top of each other to form a unified and stratified visual feeling of the space through simplified recombination. The 11-meter-high main screen test used the jade with water texture for semi-permeable treatment.

空间的主色调是柔和的米白色，配以温暖的橡木木饰面，局部镶嵌黑色金属，蓝色玻璃的装饰画非常醒目，让空间充满趣味。

“横看成岭侧成峰，远近高低各不同”，由诗词中提炼而来的山脉概念，相应简化后演变得到的门扇图腾和首层模型区的墙身木勒山脉造型，让每个空间在统一中有变化，和山居沉稳与四季变化一样，随着时移产生不同的景致，这就是设计师想要给空间营造的效果。

The main tone of the space is soft rice white and warm oak surface, local inlaid black metal, blue glass decorative painting very striking, making the space full of fun

As the ridge side into peaks, distance and height are different, the concept of the mountain from poetry, the corresponding simplified evolution of the door totem and the first model area of the wall Müller Mountain shape, so that each space in the unity of the changes, as the mountain residence calm and changes with the seasons, with the change of time, this is the effect of the designer Amus to create space.

美的檀府
MIDEA TanFu

项目地点：中国广州 / Location : Guangzhou, China
项目面积：660 平方米 / Area : 660 m²
设 计 师：陈正茂 / Designer : Cheng Zhengmao

IAI 设计优胜奖

陈正茂
Chen Zhengmao

“采菊东篱下，悠然见南山”是五柳先生的山居之趣。“蝉噪林逾静，鸟鸣山更幽”是王籍的山居之感。

项目现场依山而建，绿树成荫，建筑面积650平方米，设计师陈正茂运用山峦起伏的元素处理平面布局，运用了方圆结合的方式去表达空间节奏，让空间的收放和山峦一样有疏有密，有高有低，让人感受到不同的空间乐趣。山峦起伏的元素及璞玉元素的运用，配搭不同而有趣味性的雕塑，在灯光和材质的映衬下散发出不同空间的主题特征。由海螺演变出来的旋转楼梯，在周边瀑布墙纸围绕中旋转而上，让人在 空间中一步一景有曲径通幽的意境。

The project site is built according to the mountains, green trees, the building area is about 650 square meters, the designer Amus used the mountain undulating elements to deal with the plane layout, the designer has used a combination of means to express the spatial rhythm, so that the space is as sparse as the mountains, high and low, people feel the different spatial fun. Mountain undulating elements and elements of jade use, with different and interesting sculpture, under the light and materials of the distribution of different space theme features, by the conch evolution of the spiral staircase, in the surrounding waterfall wallpaper around the rotation of the second floor, let people step by step in the space has a smooth and quiet mood.

燕西书院接待中心
Yanxi Academy Reception Center

项目地点：中国湖南 / Location : Hunan, China
项目面积：1000 平方米 / Area : 1000 m²
设 计 师：黄婷婷、杨俊 / Designer : Huang Tingting, Yang Jun

IAI 设计优胜奖

黄婷婷 Huang Tingting
杨俊 Yang Jun

推开书院接待中心大门，天印地砚的气势极具视觉冲击力，通过墨色勾勒，素色点染，体现一股墨韵书香的气息。全透明落地玻璃窗让户外翠色全部纳入室内，让人有种丝丝清凉与静谧。以王羲之的《兰亭序》书法片段作为镂刻端景，虚透竹影，引雀停留，意境悠长。

整个设计让这不大的一方天地，有容山水、承书韵、启思辨的气度，希望能以敏锐的思维与踏实的匠心，显现中国人骨子里的诗书生活。

The Chinese stamp and ink stone at the gate adds a strong visual graphic impact. Visual design elements – form, material and color – are the basis for the visual language. Chinese traditional design has a history going back thousands of years and follows a philosophy that is unique. In China, traditional design has a deep connection with the past and plays a momentous role in forming the identity of the Chinese. Using glass floor to ceiling windows and part of Wang Xizhi' s most noted and famous work, Preface to the Poems Composed at the Orchid Pavilion, in the background explores traditional Chinese visual design elements with a goal of incorporating them into contemporary design, which is an effective way to inherit and transmit Chinese history and culture.

梵奢生活馆
Brahma Living Hall

项目地点：中国佛山 / Location : Foshan, China
项目面积：750 平方米 / Area : 750 m²
设 计 师：关升亮 / Designer : Guan Shengliang

IAI 设计优胜奖

关升亮 Guan Shengliang

致力于通过设计研究人与自然、人与空间、人与物的关系和相处之道，奉行“大道至简”的设计哲学。追寻简约、自然、艺术、人文的美学观念，作品致力于追寻光与诗意、现代与优雅的共融与对话。

Committed to the study of the relationship between human and nature, human and space, and the way of getting along with people and things through design, and pursue the design philosophy of "simplicity from the main road". Pursuing the aesthetic concepts of simplicity, nature, art and humanism, the works are devoted to the integration and dialogue of light and poetry, modernity and elegance.

梵奢家居致力于带给人们“意式极简”的生活美学和家居品味，梵奢生活馆是为了展示这一理念而设计营造。设计师在空间内外的设计运用“减法”的概念，以一种低调纯粹的姿态，传达与家具一脉相承的价值理念。在设计师营造出一个统一的高级灰的基调背景下，一件件现代、简约、个性的家具犹如艺术品一般，在空间中绽放出优雅迷人的光芒，让人难忘。

Van Luxe Home is dedicated to bringing people "minimalist Italian style" life aesthetics and home taste, Van Luxe Living Hall is designed to demonstrate this concept and build. Designers use the concept of "subtraction" in the design of space inside and outside, with a low-key attitude of purity, to convey the same value concept as furniture. Designers create a unified high-level gray tone background, a modern, simple, personalized furniture like a work of art, in the space of elegant and charming light, unforgettable.

保利春晓
Poly Spring

项目地点:中国广东 / Location : Guangdong, China
项目面积 : 285.2 平方米 / Area : 285.2 m²
设 计 师 : 王小锋 / Designer : Wang Xiaofeng

IAI 设计优胜奖

王小锋 Wang Xiaofeng

毕业于广州美术学院环境艺术设计系学士学位，从业近二十年。2008 年联合创立 SNP，任总设计师。多年的建筑结构到室内设计的丰富经验，结合他迸发的创意策划能力，为客户提供极具产品力的创意作品。

The graduated from the Guangzhou Academy of Fine Arts with a bachelor's degree in environmental art design. He has been in the industry for nearly 20 years and has led many domestic and international top-level private house designs,co-founded SNP in 2008, served as chief designer, many years of experience in building structure to interior design, combined with his creative planning ability, to provide customers with Productive creative work.

一座柏拉图式的极简主义建筑，外建筑与内建筑如阳春白雪，矗立在葱茏的树影中。建筑整体呈现180度向阳观景面，内部设计成一个无立柱空间。在这里，没有一件附加于建筑之上的多余之物，没有无中生有的变化，有的只是轻盈通透的建筑和里外流动的空间，以极简映衬繁华，构筑一个舒适放松、与现实若即若离的乌托邦。

A Plato style minimalist building, is painted in pure white inside and outside, like a dazzling white hill on a green lawn. he building as a whole shows a 180 degree view of the sun, and the interior of the building is designed to be a pure no column space. Here, there is no superfluous thing attached to the building, no unsuccessful changes, some only light and transparent buildings and the floating space inside and outside, in a very simple and prosperous way, to build a utopian which is comfortable and relaxed with reality.

北京保利和锦薇棠销售中心
Beijing Poly Belle Ville Sale Center

项目地点:中国广东 / Location :Guangdong
项目面积：1085 平方米 / Area : 1085 m²
设 计 师：王小锋 / Designer : Wang Xiaofeng

IAI 设计优胜奖

高低群山的概念无缝地嵌入在首层建筑的设计和功能布局上，阶梯式活动区选用较深的木色与灰色的柱体相搭配，宽面的阶梯可作为阅读区，作阶梯座位或陈列艺术品，呈现出前卫与实用共生的先锋理念。洽谈区沿着观景面有序分布，选用视觉与触觉感受轻松的沙发组。休闲区与VIP接待室延续了冷灰的色调，现代艺术简洁的表现形式、冷暖和谐相容是设计师对艺术生活的理解。

The concept of high and low mountains is seamlessly embedded in the design and functional layout of the first floor. The ladder type active area selects the deeper wood color and the gray column, the ladder of the wide surface can be used as the reading area, as the ladder seat or set out the art work, showing the pioneer idea of the avant-garde and the practical symbiosis. The negotiation area is orderly distributed along the viewing area, and the sofa group with visual and tactile feeling is relaxed. The concise expression of modern art, the compatibility of warmth, warmth and harmony is the designer' s understanding of artistic life.

光束的感动展演室
Moving Comes from Beam Exhibition Room

项目地点：中国台湾 / Location :Taiwan, China
项目面积：314 平方米 /Area : 314 m²
公司名称：坊华室内装修设计 / Organization Name : FH Design

IAI 设计优胜奖

以自然舒适的色调融合，运用木材、白色墙面、石板或玻璃等天然建材。透过作品、光线与艺术氛围的温馨美感，令人感觉舒适，赋予心灵净化般的疗愈效果。

强调人与自然视野的空间感受，擅长观察与光线折影是展演厅的视觉张力，在材质的运用与文化艺术上，呈现出最好的展览体验。自然流动的曲线场域，让处在这个空间的人们都享受此刻的感动。

The beauty of artworks and illuminations make people feel comfortable and pleasant by applying natural-style colors and construction materials such as woods, white walls, stone plates, glass, etc.
The design idea is derived mainly from humanity and nature.
The delicate design of illuminations brings to people wonderful experiences from the perspectives of constructions materials and artworks. Meanwhile, the live quality of the show room is of great moment for guests staying anywhere in this space.

坊华设计 FH Design

坊华室内装修设计有限公司成立于1998年，拥有 20 年以上的工程经验，对施工品质能精准掌握，在设计层面上更注重不同空间的舒适、创意及实用价值。近年来，我们积极参与各大国际设计盛事，多组设计作品荣获金、银等大奖的肯定。我们擅用天然材质装饰空间，融入艺术与文物，感受自然的呼吸与律动，享受美学与生活之间的共鸣。

Founded in 1998, Fanghua Interior Decoration Design Co., Ltd. has more than 20 years of engineering experience, and can accurately grasp the construction quality. At the design level, it pays more attention to the comfort, creativity and practical value of different spaces. In recent years, we have actively participated in major international design events, and many sets of design works have won the recognition of gold and silver awards. We use natural materials to decorate the space, blend art and artifacts, feel the natural breathing and rhythm, and enjoy the resonance between aesthetics and life.

山居
Mountain House

项目地点：中国云南 / Location : Yunnan, China
项目面积：180 平方米 /Area : 180 m²
设 计 师：李建明 / Designer :Li Jianming

IAI 设计优胜奖

我们总是借口忙碌而忘记忙碌的意义。我们总是借口为了生活而忘记生活。万物有空隙，那是光照进来的地方。心中有空隙，那是等待生活归来的地方。我总在想，心中的生活离我们到底有多远。这并不是空间意义上的距离，而是心中是否留下了等待生活归来的空隙。有了这样的空隙，也许，只要一缕阳光，就能勾起我们的共鸣。而都市人，可能更需要一个切实让生活回归的地方，需要有一个能够静下来的空间。或者说，需要这样一个空间，来打开他们心中为生活保留的那个空隙。

李建明 Li Jianming

著名雕塑艺术家
山隐（北京）设计集团总裁
山隐朴风（上海）建筑规划设计有限公司总经理
山隐朴风（云南）装饰设计有限公司总经理
DAS 大森设计顾问有限公司创始合伙人
风隐文化董事长

Famous Sculptor
Shanying(Beijing) Design Group CEO
Shanying Pufeng(Shanghai) Architectural Planning and Design Co., Ltd. General manager
SHARING(Yunnan) Decoration Design Co., Ltd. General manager
DAS Design Founding Partner
Fengying Culture Chairman

We always excuses busy and forget the meaning of busyness.We always excuse us to forget about life for life.Everything has a gap, that is where the light comes in.There is a gap in my heart, and that is where I wait for my life to return.I always wonder how far my life is from us.This is not a distance in the sense of space, but whether there is a gap in the heart waiting for the return of life.With such a gap, perhaps, as long as the sun shines, it will evoke our resonance.Urban people may need a place where life is truly returning. There is a need to have a space to calm down.Or, you need such a space to open the gap in their hearts that is reserved for life.

这样一个空间，能让你静静地与时间，与天地，与风，与鸟鸣，与自己相处。在这个空间里，阅读、喝茶、钓鱼，做点什么都好，或者，什么也不做。只是那样待着，发呆，看光影的移动。无论是否有爱人陪在身边，无论是否有一两知己一起谈笑。都只是消磨时光，不提纷扰。这样一个空间，让你自己采摘蔬菜，洗手做羹，吃简单的饭食，体验难得的人间清欢。

山居，是我的设计作品中，思考最多的一个。城市纷扰，追逐太多，总需要有一个地方，提醒我们，留给生活归来的空隙，这样，每天的奋斗才都能被阳光照亮。

Such a space allows you to quietly interact with time, with the world, with the wind, with the birds, and with yourself.In this space, read, drink tea, fish, do something better, or do nothing.Just stay that way, daze, watch the movement of light and shadow.Whether or not there is a lover to accompany you, whether or not there are one or two confidants talking and laughing.It’ s just a matter of killing time, not to mention trouble.Such a space allows you to pick vegetables yourself, wash your hands and make sputum, eat simple meals, and experience rare human love.Shanju is one of the most thought-provoking of my design works.The city is confusing, chasing too much, there is always a need to have a place to remind us to leave the gap of life back, so that every day's struggle can be illuminated by the sun.

Office Space

室内 —— 办公空间

静界
Sukhavati

项目地点：中国福建 / Location :Fujian,China
项目面积：240 平方米 /Area : 240 m²
设 计 师：蔡天保，张建武 / Designer : Cai Tianbao , Zhang Jianwu

IAI 最佳设计大奖

张健武 & 蔡天保
Zhang Jianwu & Cai Tianbao

2019 德国 IF 设计奖
2018 意大利 A' 设计大赛银奖
2018 德国红点奖的红点设计大奖
2018 第十一届台湾室内设计大奖的商业空间 / 休闲空间 TID 奖
2017 中国设计品牌大会最具投资价值商业品牌空间
2017 第八届中国国际空间设计大赛商业、金融、售楼处空间 / 工程年度创新作品奖

2019 German IF Design Award
2018 Italy A'Design Award Silver Award
2018 German Red Dot Award Red DOT Design Award
2018 The 11th Taiwan Interior Design Award Commercial Space/Leisure Space TID
2017 Award
China Design Brand Conference, the most investment-worthy commercial brand space
2017 The 8th China International Space Design Competition Commercial, Finance,
Sales Office Space/Engineering Annual Innovation Works Award

拾阶而上，映入眼帘的是一头金色的熊趴在玻璃上向内窥望，这是致敬艺术家劳伦斯·爱勋在美国丹佛会议中心创作的蓝熊，它的趣味性打破了走廊的单调。两条光线分别指引着两个不同区域的入口。左边是设计部，清新雅致的用色与选材都能让心马上平静下来；设计师提升了办公区域的地面，上空悬吊着可根据功能需求、调节灯光色温和亮度的灯具，将下面作为储物间和设备间，既丰富了空间的层次，又解决了收纳的问题。右边是由三个独立空间打通而成的、完全通透纯白的空间，视觉中心是多功能的木盒子，盒子的围挡采用矩管格栅的形式，通过格栅观看楼梯边的佛像及楼梯下行走的人，会让平静的空间产生互动的波澜。

Up the stairs, a golden bear on the glass is peeping up on the glass. This is the blue bear created by Lawrence Argent in Denver Convention Center in the United States. It is interesting to break the monotony of the corridor. The two light guides the entrance of two different regions. The left is the design department, designers raised the floor of the office area, hanging over the lamps and lanterns which can adjust the light color temperature and brightness according to the functional requirements. The following will be used as the storage room and the equipment, not only rich in the room and the equipment. The level of space, and the problem of acceptance. On the right is a completely white space made up of three separate spaces, and the center of vision is a multifunctional wooden box, the enclosure of which is in the form of a rectangular tube grille.

SLT 办公室
SLT Office Design

项目地点：中国北京 / Location : Beijing , China
项目面积：157 平方米 /Area : 157 m²
公司名称：北京艾奕空间设计有限公司 / Organization Name : AE Architects Design

IAI 杰出设计大奖

艾奕设计
AE ARCHITECTS

艾奕空间设计创立于北京，得益于创始人 Abel Erazo 在建筑设计、景观设计以及室内设计领域二十多年的实践经历。我们的艺术哲学观受古代东方禅宗和西方现代化摩登的影响，形成了简约、朴素、纯洁、象征主义的风格，自成一派。我们拥有丰富的餐饮、办公、酒店、住房、商业和文化设计经验，确保创造高效、舒适和独特的空间。

AE Architects was founded in Beijing and establish its practice from the 20 years of experience of Abel Erazo in Architecture, Landscape and Interior Design. Our philosophy has been influenced by Ancient Asian Zen and Western Modern. Taking from both the ideas of simplicity, pureness and symbolism. With vast experience on food and beverage, office, hospitality, housing, commercial and cultural design, we ensure the creation of efficient, comfortable and unique spaces.

SLT办公室是一系列时尚品牌的总部，该品牌有六百多家连锁店遍布全国各地。此次的设计任务不仅仅是为公司提供办公空间，同时还需要有展示的功能以及接待国际客户的空间。整个设计中采用温暖的色调、亲和的环保材料，以及绿色的自然元素，充分诠释集团的哲学理念，专注年轻一代，关注未来的生活方式。我们想要表达的是城市生活与自然相互联系所达到的平衡。优雅与简约在空间中以一种舒适和交互的形式呈现。办公室坐落于北京三里屯时尚区的中心，平面布局也结合了三里屯SOHO的几何构成。

SLT office is the headquarters of the fashion brand with more than six hundreds shops across the country. The commission not only include the working space for the company but also exhibitions and reception areas for international guests. Using warm tones, environmental friendly materials, green and natural elements, the design express the philosophy of the group focus on the youth and the future lifestyle. We want to express the city life connected with nature in a calm balance. Elegance and simplicity come to the space in a comfortable and interactive way. Located in the heart of the fashion district of Sanlitun in Beijing, the plan adapt its composition to the geometries of Sanlitun SOHO complex.

在入口处呈现了一面展览墙，展示的是公司的产品。木质线条元素给空间增添了几分灵动和深邃；而在前台的背景墙，石头和艺术品在工业风的外壳下相融合，给空间带来了强烈的反差效果。在平面布局的中心，是客户公共洽谈区，它共享了展示墙。

弧形边界让我们以一种年轻、充满趣味的方式打造休息和聚集区。窗户边的坐席式台阶，既可以在这享受窗外的城市美景，又赋予了公共活动区满满的活力。

From the entrance an exhibition wall display some of the products of the company. wooden linear elements add vibration and depth to the space while at the background of the reception, stone and art pieces connect with the industrial envelope and bring contrast. In the middle of the plan a guest open room share the display wall.
Round borders allow us to set the resting and gathering areas in a young and playful way. A sitting stairs by the window enjoy the views of the city and also add dynamism to the public activities.

四面围合的会议室看似简约，实则融入了技术性及实用性的细节。多孔的金属天花板不仅加强了声音吸纳功能，而且为天花板的维护提供便利。延续天花板的灵动性，圆点图案的玻璃墙赋予会议室私密性。

在平面布局的后端，私密办公室和开放工作空间共享着一面展示墙，展示着衣服及包包等产品。游走于绿植和温暖的木质家具之间，整体给人的感觉是舒适惬意、时尚潮流。此外舒适的办公空间与时尚行业也很好地搭配在一起。

The four surrounding conference rooms seem simple, but they incorporate technical and practical details. The perforated metal ceiling not only enhances the sound absorption function, but also facilitates the maintenance of the ceiling. Continuing the agility of the ceiling, the glass wall with a dot pattern gives the meeting room privacy

At the back of the plan the private offices and open working spaces share their functions with display walls of clothes and bags. Between the green plants and the warm wood the overall experience is cosy, trendy and fashionable. Matching together the comfort of the working space with the fashion industry.

TZ 办公室
TZ Office Design

项目地点：中国北京 / Location : Beijing, China
项目面积：400 平方米 /Area : 400 m²
公司名称：北京艾奕空间设计有限公司 / Organization Name : AE Architects Design

IAI 设计优胜奖

艾奕设计
AE ARCHITECTS

艾奕空间设计创立于北京，得益于创始人 Abel Erazo 在建筑设计、景观设计以及室内设计领域二十多年的实践经历。我们的艺术哲学观受古代东方禅宗和西方现代化摩登的影响，形成了简约、朴素、纯洁、象征主义的风格，自成一派。我们拥有丰富的餐饮、办公、酒店、住房、商业和文化设计经验，确保创造高效、舒适和独特的空间。

AE Architects was founded in Beijing and establish its practice from the 20 years of experience of Abel Erazo in Architecture, Landscape and Interior Design. Our philosophy has been influenced by Ancient Asian Zen and Western Modern. Taking from both the ideas of simplicity, pureness and symbolism. With vast experience on food and beverage, office, hospitality, housing, commercial and cultural design, we ensure the creation of efficient, comfortable and unique spaces.

本项目位于中国北京通州新区，是工业用地被大规模改造成写字楼和商业园区的第一步。占地40000平方米的旧厂房在过去曾属于北京印刷厂，现在已经空置，整体的全新规划不仅包括办公空间，还增添了餐厅、咖啡厅、礼堂、户外活动空间等服务。为了吸引新晋公司、创业人群和年轻一族，这个样板办公室创造出一个充满趣味的工作环境，在此可以尽情互动、聚集、交流、共享和娱乐。此次的设计思路也尊重现有的建筑结构，使用轻盈的木材、自然的绿植和清新的水，强烈地衬托出工业风裸露着的混凝土内部。我们不想干扰或破坏原来的建筑外围，所以用更当代的手法使其脱离之前曾有的建筑样式，而变为通高的两层阁楼全新空间结构。轻薄钢柱和轻质墙体共同见证新的空间功能诞生。亲近自然的元素把人与人类起源的畅想联系起来，创造出非常独特奇妙的氛围，犹如置身城市中央的绿洲。

Located in the new district of Tong Zhou in Beijing, this project is the first step of a large industrial conversion in to office and business park. The old factories, property of the Beijing printing company in the past now empty, occupy an area of 40,000 sqm. The new master plan include not only office spaces but also restaurants, cafes, auditorium, outdoor events space among other services.In order to attract new companies, start up and young users, this model office create a playful working environment in which, interaction, gather, exchange, entertain and fun are part of the daily working activities. We didn' t want to interfere or destroy the original envelope, so the new structure with its double height loft space is detached from the previous one in a contemporary way. Slim steel columns and light walls allow the new functions take place.?We didn' t want to interfere or destroy the original envelope, so the new structure with its double height loft space is detached from the previous one in a contemporary way. Slim steel columns and light walls allow the new functions take place. The natural elements connect the human with its origins and create a very unique atmosphere like an oasis in the middle of the city.

鼎天办公室
Dingtian New Office

项目地点：中国福州 / Location : Fuzhou, China
项目面积：1000 平方米 /Area : 1000 m²
设 计 师：黄婷婷 / Designer : Huang Tingting

IAI 最佳设计大奖

本案作为一个装饰设计企业创意聚集的办公空间，设计师充分地结合了自然元素和人文办公的需求点，采用完全开放的思路和丰富的空间动线设计手法，淡雅的色调使整体空间更为祥和安静，让人身心格外的放松。

在空间处理上运用纯粹的线条对整体空间进行分割，大面积的面板、整面玻璃、木栅格的应用，色调统一，和谐的过渡让空间若即若离，零散开放但又不失整体性。

As a creative office space gathered by decorative design enterprises, the designer fully combines the natural elements and the needs of humanistic office, adopts completely open thinking and abundant space dynamic line design techniques, and the elegant tone makes the whole space more peaceful and quiet, which makes people more relaxed physically and mentally.
In space processing, the use of pure lines to divide the whole space, the application of large panels, whole glass, wooden grids, uniform tone, harmonious transition to make the space if not separated, scattered and open, but without losing the integrity.

黄婷婷 Huang Tingting

中国建筑装饰协会高级室内建筑师
福建鼎天装饰工程有限公司创始人、总设计师
擅长项目：教育机构、办公空间、展示空间
设计理念：设计是给予项目最适合的展示环境及空间舒适的体验

Senior Interior Architect of China Building Decoration Association
Founder of FuJian Dian Tian Decoration Corporation and Leading designer
Specialize in design of educational institution, office space and display space of shopping centre
Concept innovation: create an atmosphere to be simple and comfortable

设计师在入口和通道都采用大面积的云朵拉灰大理石材作为装饰，略显粗犷的表面和水泥墙形成有机整体，在灯光的映照下，使空间显得深邃而富有神秘的韵味。乳化超白透明玻璃和木栅组成的背景墙，在点光和面光的相互辉映下，悠悠美景真实而又虚幻。手工编织地毯、皮质沙发和大理石茶几，空间的层次丰富而丰满，让人仿如置身美景，惬意盎然。洽谈区用整面的玻璃让空间融入周围，在保证私密性的同时又不显局促。为了延续空间的统一风格，主体办公区在布局上开放且井然有序，公共区域的座椅以白、绿为主色调，醒目而又不失协调。

Designers adopt large areas of cloud-pulled lime marble as decoration in the entrance and passageway, slightly showing rough surface and cement wall to form an organic whole, under the light, making the space appear deep and mysterious charm. Emulsified super-white transparent glass and wooden grille constitute the background wall, reflecting the mutual brilliance of spot light and face light, the beautiful scenery is real and illusory. Hand-woven carpets, leather sofas and marble tables, the level of space is rich and full, let you feel as if you are in the beautiful scenery, comfortable! The whole glass is used in the negotiation area to make the space merge into the surrounding area, which ensures privacy without being constrained. In order to continue the unified style of space, the main office area is open and orderly in layout. The chairs in the public area are mainly white and green, which are eye-catching and coordinated.

融侨大梦
Diamondream Rongqiao

项目地点：中国福州 / Location : Fuzhou, China
项目面积：600 平方米 /Area : 600 m²
设 计 师：黄婷婷 / Designer : Huang Tingting

IAI 设计优胜奖

融侨大梦位于福州闽江北CBD融侨中心ART MALL艺术商业空间，坐落在现代摩登写字群楼之间，在触手可及的地方，为精英生活带来奇妙的阅读体验。

翻开书屋入口，一股恰到好处的工业风氛围空间席卷而来。设计师将橡木建材、金属铁件元素与抛光混凝土地板结合在一起。造型简约充满现代工业感但不冰冷。没有太多的色彩堆积，棕色的钢板、橡木材质书架桌椅充满了低调沉稳的气质，触感力求质朴原始。这种返璞归真的质感，能够令人分秒之间沉浸在这个文艺的空间。线条感利落的吊线顶灯铁艺摆饰，华丽之下没有一丝赘余的元素。

Diamonddream Rongqiao is located in the Art Mall Art commercial space of rongqiao center in the northern CBD of minjiang river, fuzhou. It is located among contemporary modern offce building.

Open the entrance of the library, a proper industrial wind atmosphere space. The designer combines oak and metal elements with polished concrete floors. Simple modeling full of modern industrial feeling but not cold. There is not too much color accumulation, brown steel plate, oak material bookshelf table and chair is full of low-key calm temperament, tactility strives for simplicity primitive. This kind of uncut jade to return true simple sense, can make you immerse in the space of this artistic feature in a minute. The condole line that feels agile and agile is acted the role of art of ceiling lamp iron art is acted the role of, there is not a redundant element below luxuriant.

ET. 设计工作室
ET. Design Office

项目地点：中国澳门 / Location : Macao，China
项目面积：4000 平方米 / Area : 4000 m^2
设计师 / Designer : Eric Tam

IAI 设计优胜奖

Eric Tam

从事室内设计行业已经超过 20 年。拥有一家属于自己的设计公司一直是 Eric Tam 的梦想，他果断放弃了很好的工作机会移居澳门创立了昊设计工程公司。经过多年的努力，Eric Tam 及他的昊设计硕果累累，不仅获得许多奖项还加入了不同的协会，备受各界人士的认可。

Eric has been involved in the Interior Design Industry for over two decades. Eric always had a desire to open his own firm, he eventually left the prior company to pursue and eventually establish ET Design & Build ltd. He and ET Design have received numerous awards and joined different kind of associations.

我们希望设计一个与传统隔间布局完全不同的办公空间，这家室内设计公司专注于工作环境，充满乐趣、创意和舒适的开放性。建立这样一个工作场所是为员工提供舒适、缓解压力的工作环境，并激励员工想要在办公室工作。4000平方英尺的空间充满了游戏室、电视室、台球桌和温馨的开放式厨房。健身房也随时让工人保持身心健康。除了有趣的区域，还有2个开放式工作区、2个开放式会议区、1个开放阅读区、1个物料供应展示区、2个私人会议室和4个私人办公室。

We want to design an office space that is nothing like the traditional cubicle layout, this interior design company focus on the working environment to be fun, creative and comfortable with openness. The purposes of making such a workplace is to give employees a comfortable, yet stress relief environment to do work and also motivate employees to work in the office in the office. The 4000-square-feet space filled with game room, TV room, pool table and a homey open kitchen. A gym is also at hand to keep the workers in healthy body and mind. Beside the fun areas, there are 2 open workspaces, 2 open meeting areas, an open reading area, a material supply display area, 2 private meeting rooms and 4 private offices.

非黑
White Soul

项目地点：中国福州 / Location : Fuzhou, China
项目面积：93 平方米 /Area : 93 m²
设 计 师：敖静 / Designer : Ao Jane

IAI 设计优胜奖

敖静 Ao Jane

宜空间设计创始人，拥有十余年室内设计从业经验的资深设计师。
主导设计项目，包括地产样板间、商业、住宅等等。她主张设计能够帮助人们更好的理解生活，发现生活中你不曾在意的细节、故事。

Jane Ao, created the YI space design. Leading design projects, senior designer with more than ten years of experience in interior design including real estate showrooms, commercial, residential and more. She advocates that design can help people better understand life, discover details and stories that you didn't care about in life.

这里是一个设计工作室，只有93平方米。我们希望塑造一个沉浸式的空间。进入、被包围，然后卸下伪装与防备，寻回最放松的状态去创造。于是，纯粹的白是这个氛围的低衬。

入户的方向反转，是进入空间的第一步。在这里脱鞋、步入，我们让这样的行为，成为一种意识上的卸载，卸下防备与伪装。进入空间，T台一般的走道，以斜置的格局，破除了原本规矩的方正。白、简单、纯粹、舒适，是这个空间的主题。

Here is a design studio, only 93 ㎡ . We want to create an immersive space which makes us be surrounded and embedded in, then be rid of disguise and defense, reaching to the most relaxed state for better creation.

After stepping into it, one is ushered into a reversal.Taking off shoes here and then walking in, suggest taking off one' s defense and disguise. Into the space, the tilted aisle, like a catwalk, gets rid of the conventional rules.
White,simple, pure and comfort, is the theme of this space.

纽约梦想
New York Dream

项目地点：中国台湾 / Location :Taiwan , China
项目面积：44 平方米 / Area : 44 m²
公司名称：创空间集团 / Organization Name : Creative Group

IAI 设计优胜奖

創空間
CREATIVE GROUP

创空间 CREATIVE GROUP

汇集建筑与室内设计背景的多元人才，配合组织化管理模式，将工程进度做系统化的全面控管，
提供从前期沟通、设计、合约拟定到后期执行的完整服务。2000 年成立“权释设计”。2014 年、2015 年陆续成立“CONCEPT 北欧建筑”与“JA 建筑旅人”建筑设计公司，传递人与自然和谐共生的理念精神，秉持着“提升国人生活品质、共创生活美学体验”为目标持续优化。

Bringing together diverse talents in the background of architecture and interior design, and coordinating the organizational management model, systematically and comprehensively control the progress of the project, providing complete services from pre-communication, design, contract formulation to post-execution. In 2000, the company established the "ALLNESS Design" . In 2015, it established the "CONCEPT " and the "Journey Architecture" architectural design company to convey the spirit of harmony between Humanity and Nature Environment, and uphold the "quality of life of the people." is the continuously optimized goal .

破除老屋高度限制及采光问题，运用玻璃的穿透性将空间的隔阂化为无形，并将走道重整于空间中轴线，让人流穿梭于空间而无局促感，因人而生的热情能量也得以流动在空间各角落。象征纽约地铁的红色在天花板画出一道红色轨道，承载着过去的记忆也启动未来的梦想。教室试图让视觉聚焦于人所传达的情感与力量，去除色彩干扰，让使用者在空间能专注于自己的训练，全心挥洒；在这个承载梦想舞台的空间里，彼此交流与对话，舞动永不止息。

The Breaking the height restriction of the old house and the problem of lighting, using the penetration of the glass to transform the space into invisible, and reorganizing the aisle in the central axis of the space, letting people flow through the space without a sense of urgency, the enthusiasm of life also flows in every corner of the space. Drawing a red track on the sky in the red symbolizing the New York subway, carrying the memories of the past also launches the dream of the future. The classroom attempts to focus the attention on the emotions and strengths conveyed by the people, remove the color interference, let the users focus on their own training in the space, and devote themselves wholeheartedly; in this space carrying the dream stage, communicate and dialogue with each other, dance forever non-stop.

净化
Purification

项目地点：中国江苏 / Location : Jiangsu, China
项目面积：600 平方米 / Area : 600 m²
设 计 师： 董则锋 / Designer : Dong Zefeng

IAI 设计优胜奖

董则锋 Dong Zefeng

中国建筑装饰协会高级室内建筑师、高级住宅室内设计师，CIID 中国建筑学会室内设计分会会员。1996 年毕业于南京艺术学院装潢设计专业，擅长欧式、新古典、美式、中式风格。

美国 Best of Year Awards 2018 年度最佳设计大奖
亚太设计大赛商业空间钻石奖
亚太设计大赛办公空间至尊奖
华语设计领袖榜 2018 年度 100 位卓越设计人物

Senior Interior Architect, Senior Residential Interior Designer of China Building Decoration Association, member of CIID China Architecture Society Interior Design Branch. In 1996, he graduated from Nanjing Art College with a major in decoration design. He is good at European, Neoclassical, American and Chinese styles.
US Best of Year Awards 2018 Design of the Year Award
Asia Pacific Design Competition - Commercial Space - Diamond Award
Asia Pacific Design Competition - Office Space - Supreme Award
Chinese Design Leaders List of 100 outstanding design characters in 2018

净化是一个办公室设计项目，位于中国江苏省苏州市张家港。一楼以白色为主，中置吸水石及观溪流，溪声阵阵，赏心悦目。二楼电梯厅东侧为中式斗拱大理石柱，西侧为欧式爱奥尼柱，象征东西文化的交流与碰撞。二楼的金属管道隔断为旧式自来水管弯接成“THINKER”字样，寓意一为财富管道，二为只有常思，思想才不会生锈。主体墙面为家具厂废弃木料回收利用，倡导绿色环保装修的理念。顶面为轻钢龙骨附龙骨密排，与镀锌水管相映成趣。

Purification is an office design project in Zhangjiagang, Suzhou, Jiangsu, China. The first floor is dominated by white, with the middle water absorbing stone and the stream of view, pleasing to the eye. The eastern side of the elevator hall on the second floor is Chinese arched marble pillars, and the western side is European Ioni pillars, which symbolize the exchange and collision of eastern and Western cultures. The metal pipe partition on the second floor used old tap water pipe to bent the word "THINKER" . The moral is that on the one hand, it represents the channel of wealth, and on the other hand, it is only by thinking often that ideas do not rust. The main wall is the recycling of waste wood in furniture factories, advocating the concept of green environmental protection and decoration. The light steel keel is used on the top surface to match the galvanized water pipe.

HOLLYS COFFEE

叙品新疆办公室
Office of Xinjiang Sin Kiang Branch of Xu Pin Design

项目地点：中国新疆 / Location : Xinjiang, China
项目面积：2000 平方米 / Area : 2000 m²
设 计 师：蒋国兴 / Designer : Jiang Guoxing

IAI 设计优胜奖

蒋国兴 Jiang Guoxing

叙品空间设计有限公司董事长兼设计总监，他 1996 年毕业于厦门工艺美院。从业二十余年，他始终坚持原创设计，并将禅文化运用到空间设计，从而使空间更有深度。凭借对原创的坚持和对东方文化的理解，蒋先生获得了很多国内外的设计荣誉。如英国安德鲁马丁，意大利 A' 设计大奖，德国 IF 设计奖等。

Xupin Space Design Limited Company's Chairman&Design Director,He graduated from Xiamen institute of arts and crafts in 1996 and has been working as a designer for more than 20 years.He always stick to original design ,meanwhile he apply Zen to space design which make the space more thoughtful and appealing.With the persistence of original design, the understanding of Oriental culture,Mr Jiang won a lot of honners in both home and abroad. Such as Andrew Martin Interior Design,Italian A' Design Award ,Gremany IF Design Award and so on.

整个空间以白色为基调，局部用蓝、黄、绿色做点缀，以有“色”计为设计思路，综合使用原木、玻璃、石材、镜面等多种材质，以简洁有力的手法，打造出富于变幻充满想象的多元化空间。走道是空间的点睛之笔，彩色玻璃重叠交错，亦真亦幻。富有张力的雕塑贯穿其中，与空间相得益彰。从飞天的猪、一脸陶醉的马，到桀骜不逊的人物，喜怒哀乐众生百态……是奔腾的艺术，更是无声的乐章。这里是办公室，更是艺术展厅，雕塑作品定期更新，为参观客人带来新鲜的视觉体验。通透明亮的开放式办公空间，星空吊顶灵感源于褶皱的纸张，每个出色的作品背后，都凝聚了叙品设计团队的智慧和心血，褶皱了无数张纸，历经无数个披星戴月的夜晚……年轻的设计师们在这里尽情释放工作热情，激发创作灵感。墙面山料肌理质感，灯光明暗变幻和不同色调过渡，带来震撼的视觉感受。这里也可以作为艺术展览、时尚人群聚集地，多功能开放属性赋予空间更多可能。虽居喧嚣闹市，却有阳光、绿植、星空相伴，为你诠释另类回归自然。置身其中，倾听人与自然的对话、艺术与生活的交响。

The whole space is white as the keynote, partly embellished with blue, yellow and green, with "color" design as the design idea, the comprehensive use of logs, glass, stone, mirrors and other materials, with concise and powerful methods, to create a versatile space full of changeable imagination. The corridor is the focal point of the space. The overlapping and staggering of stained glass is also true and illusory. Tension-rich sculpture runs through it, complementing each other with space. From the flying pigs, the intoxicated horses, to the unruly characters, emotions,happiness, anger, sorrow and joy, beings all shapes and sizes. It's the art of Pentium, but also the silent movement. This is the office, but also the art exhibition hall, and the sculpture works are updated regularly, bringing fresh visual experience to visitors. Transparent and bright open office space, Star ceiling inspired by folded paper, behind each excellent work, has condensed the wisdom and efforts of the narrative design team, folded countless sheets of paper, through countless starry night... Young designers are here to release their enthusiasm and inspire their creativity. The texture of the wall materials, the changing light and shade and the transition of different tones bring a shocking visual feeling. This is not only an office, but also a gathering place for art exhibition and fashion crowd. The multi-functional and open attributes give space more possibilities. Although living in a noisy city, there are sunshine, green plants and stars accompanying you, interpreting the alternative return to nature. In it, listen to the dialogue between man and nature, the symphony of art and life 。

伯明翰足球俱乐部香港办公室
Birmingham Football Club Hong Kong Office

项目地点：中国香港 / Location：Hong Kong, China
项目面积：2100 平方米 / Area：2100 m²
设 计 师：林志辉 / Designer：Leslie Lam

IAI 设计优胜奖

林志辉 Leslie Lam

林志辉设计事务所设计总监，毕业于中国香港室内设计专业学院。30年设计生涯里本着"轻装修、重装饰"的设计理念，再融入国际性时尚设计元素，至今成功打造了三百多个不同风格的项目。由2016 年至今获得多项设计奖，设计作品亦获多家出版社发表刊登。他坚信创作之路永无止境，并且超越自我，在未来创造更多有特色的作品。

Lin Zhi Hui Design Director of Lam Chi Fai Design, graduated from the Hong Kong Institute Of Interior Design. In the 30-year design career, the design concept of "light decoration, heavy decoration", and then integrated into international fashion design elements, has successfully created more than 300 different styles of projects. A number of design awards have been won since 2016, and design works have also been published by various publishers.He firmly believes that the road to creation is endless and transcends himself, creating more distinctive works in the future.

本案例共分为 2 层，总面积约两千一百平方米，以现代港式为设计风格，整体采用冷暖色调搭配，既保持了商业空间的沉稳大气，又不会显得过于冰冷。从平面规划开始通过几何弧形划分区域，加入智能化灯光系统让空间有不一样的体验；同时利用大量不规则的线条营造出动感和时尚的一面，令整体空间生动起来；另外办公家具均来自奥地利 Bene、德国 Sedus 和 Walter Knoll 等进口高端家具品牌，与整体空间融为一体，突显格调与品味。

With the modern port style as the design style, the overall use of warm and cold color matching, not only to maintain the calm atmosphere of commercial space, but also not too cold. Starting from the plane planning, we can divide the area by geometric arc and add intelligent lighting system to make the space have different experience; at the same time, we use a large number of irregular lines to create a dynamic and fashionable side, which makes the whole space vivid; in addition, office furniture comes from imported high-end furniture brands such as Bene, Sedus and Walter Knoll of Austria, which merges into the whole space. Show style and taste.

生长的盒子
Growing Box

项目地点：中国广州 / Location :Guangzhou , China
项目面积：1300 平方米 /Area : 1300 m^2
设 计 师：刘国海 / Designer : Liu Guohai

IAI 设计优胜奖

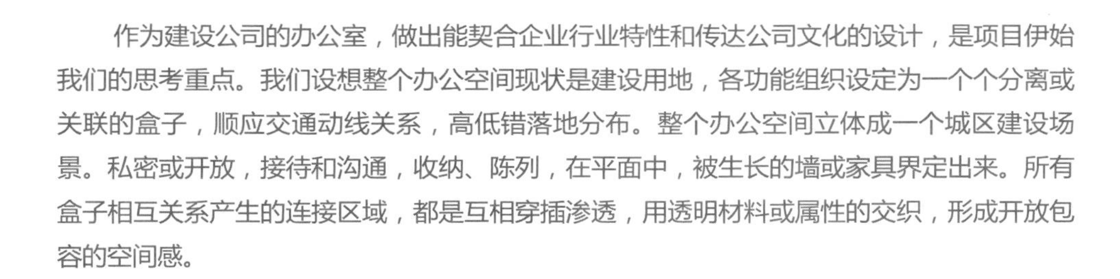

作为建设公司的办公室，做出能契合企业行业特性和传达公司文化的设计，是项目伊始我们的思考重点。我们设想整个办公空间现状是建设用地，各功能组织设定为一个个分离或关联的盒子，顺应交通动线关系，高低错落地分布。整个办公空间立体成一个城区建设场景。私密或开放，接待和沟通，收纳、陈列，在平面中，被生长的墙或家具界定出来。所有盒子相互关系产生的连接区域，都是互相穿插渗透，用透明材料或属性的交织，形成开放包容的空间感。

At the very beginning of the project, for the office room of the construction company, ours focus is to make the design that match the characteristics of the enterprise industry and convey the culture of the company. We envisage that the entire office space is construction yard, and each functional room as a separate and associated box, which conforms to the traffic relationship by high and low. The entire office space is a urban construction scene in three-dimensional. Private or open, reception and communication, storage, display were defined by the wall or furniture in the plane. The connection areas of all the boxes are interpenetrated, were intertwined with transparent materials, to form an open and inclusive space.

刘国海　Liu Guohai

毕业于华南理工大学，2001 年开始从事建筑及室内设计行业，2014 年作创办刘国海建筑设计事务所至今。在不同设计领域中不断获得尝试和实践，在建筑、园林、vi 系统等也多有涉猎。作品陆续收录于《中国手绘设计 40 人》《经典设计方案》《顶级办公空间设计》《亚太室内设计双年奖》《艾特奖》《APDC》等书籍、设计杂志及赛事之中。

Graduated from South China University of Technology,Started construction and interior design industry in 2001,In 2014, Liu Guohai Architects was founded.Continuous trial and practice in different design areas,In architecture, gardens, vi systems, etc. are also involved. His works have been included in "China's Hand-painted Design 40", "Classic Design", "Top Office Space Design", "Asia Pacific Interior Design Biennial Award", "Aite Award", "APDC" and other books, design magazines and events. in.

IDDW设计办公室
IDDW Design Office

项目地点：中国南京 / Location : Nanjing, China
项目面积：93 平方米 /Area : 93m²
设 计 师：王伟 / Designer : Wang Wei

IAI 设计优胜奖

王伟 Wang Wei

道伟（IDDW）室内设计有限公司
高级室内建筑师、建筑摄影师
CIID 中国建筑学会室内设计分会会员
AIDIA 亚洲室内设计联合会专业代表
APDF 亚太设计师联盟会员
IFI 国际室内建筑师联盟会员
JSIID 江苏省优秀室内设计师称号
CIID 中国室内官方推荐设计师
中国地区 50 位优秀青年室内设计师
大师对话营推荐青年设计师

IDDW Interior Design Co., Ltd.
Senior Interior Architect and Architectural Photographer
CIID China Institute of Interior Design Branch Member
Professional Representative of AIDIA Asian Interior Design Federation
APDF Asia Pacific Designers Association Member
IFI International Interior Architects Alliance Member
JSIID JiangSu Province Excellent Interior Designer Title
CIID China Indoor Official Recommendation Designer
50 Excellent Young Interior Designers in China
Master Dialogue Camp recom-mends young designers

本案为IDDW道伟设计南京办公室，是一套96平方米的LOFT空间。高度5米，单面采光。设计师为了更好地解决空间高度及采光问题，在门厅处采用天然透光石墙面，并营造出了与传统办公室不同的氛围。

天然石材的肌理与纹路，赋予空间以经得起时间打磨的持久品质。设计师在色彩和造型的趋简中寄寓至上的创造力，在深调的基础上搭配定制的软装，使充满现代感的优雅流淌在休憩区。

空间在材质的催化下传达含蓄而克制的气场，设计师通过概括提炼与抽象联系的手法，将折线、折面与体块的形态之美表达得淋漓尽致。

This case is designed for IDDW Daowei Nanjing Office, which is a 96 square meters LOFT space. Height 5 meters, one-sided lighting. In order to better solve the problem of space height and lighting, designers use natural transparent stone wall in the hall, and create a different atmosphere from traditional office.
The texture of natural stone endow space with lasting quality that can withstand time polishing. Designers embody supreme creativity in the simplification of colors and shapes. On the basis of deep tone, they mix with customized soft clothes to make the elegance full of modernity flow in the recreation area.
Space conveys implicit and restrained atmosphere under the catalysis of material. Designers express the beauty of the shape of broken lines, folded surfaces and blocks vividly by means of generalization and abstraction.

梦想之邸
House Of Dream

项目地点：中国台湾 / Location : Taiwan, China
项目面积：47 平方米 /Area : 47m²
公司名称：宸舍室内装修设计有限公司 / Organization Name : Chenshe Interior Design

IAI 设计优胜奖

宸舍设计
ChenShe Interior Design

空间因为人而具有生命力、而设计理应秉持着以人为本的理念、以舒适与机能为空间的核心。
宸舍设计创立于 2007 年，从事之设计领域涵盖住家空间、商业空间、办公空间、大型聚会场所与建筑设计，创立十年以实作经验实践想像中的空间，以使用人为设计出发点。在人与美感和机能当中取得平衡，让使用者在空间里能够感受到空间美感所带来身心灵的愉快以及感动。

Space has vitality because of people, and design should adhere to the people-oriented concept and the core of comfort and function.
Chen She Design was established in 2007, the business department and the engineering department. Chen She Design fields cover residential space, commercial space, office space and large gatherings. Site and architectural design, created a decade of practical experience to practice the space in the imagination, to use the artificial design starting point
Balance between people and beauty and function
let the user feel the pleasure and touch of the body and mind brought by the space beauty in the space.
Place a few simple chairs with small styles and small single tables, so that everyone can place combinations and directions in a free and relaxed situation, or one or more people, so that there are changes in the negotiation.

梦邸之乡是一打造设计的团队办公处所，这是一个打造梦想的地方，打造出梦想中生活气质的感知，每个案子的规划背后，都有一个摇篮操作者实现空间的力量，力量就来自梦想之邸这个原始空间，给予设计团队无限的全知灵感，不断实现客户的蓝图愿景。

"House of dream," the office for an interior design team, is for dream fulfillments, and for life realizations. Every project comes from the countless efforts originated from "House of dream" that inspires the design team to fulfill the clients' dreams.

MAX 住宅
MAX House

项目地点：中国广州 / Location：Guangzhou, China
项目面积：375 平方米 /Area :375 m²
设 计 师：谢法新 / Designer：Xie Faxin

IAI 设计优胜奖

谢法新 Xie Faxin

代表作：
MAX HOUSE "木" 之所及
设计理念：
创作上，总是有着自我的信仰与期待，注重寻找潜在因素，反而更独树一格。从中挖掘出独特的空间秩序，不断寻找、探索、尝试新的可表现手法，力图追求设计的特性和个性。

Magnum opus
MAX HOUSE,WOODEN INSTITUTE

Concept of Design
About creation, I always have my own belief and expectation to focus on finding potential factors and make more unique. In addition, I commit to unearth a unique spatial order. Through constantly seeking, exploring, and try new ways to express, strive to pursue the characteristics and personality of design.

本案设计始于通过空间的延续、重组、再现。正如以表述空间与使用者的关系来回应建筑本质，设计以多样探索赋予空间更大的想象力，如同音乐旋律，它也有高低起伏的变化。本案尝试加入艺术元素，编织出属于这个空间的独有乐章，让生活就像游于艺廊般，享有被心动事物环抱的幸福感受，这是我们对于办公空间的一种新诠释方式。

每个体块看似连贯，却又互不制约。同时，让每个空间都具备其该有的功能形式。整体区域开敞舒适，关注点在人与空间关系的探索，我们希望这种形式感能延续到设计师工作的状态中，开放心态，敢于走出每一步。这也是我们希望呈现给每个来访者的空间触感。

The design starts from the continuation, reappear and reproduction of space. Just as space expresses its relationship with the user to respond to the nature of the architecture, the design gives space a greater imagination by various explorations, like the melody of the music, and it also has high and low fluctuations. This case attempts to join the art element, weave the unique movement of this space, making life like a gallery like, enjoying the happy feelings embraced by the heart, which is a new way of interpretation of our office space.

Each block appears to be coherent, but it does not restrict each other. At the same time, each space has its own functional form. The whole area is open and comfortable, focusing on the exploration of the relationship between people and space. We hope that this sense of form can continue into the state of the designer's work, open mind and dare to walk out of every step; it is also the space touch that we wish to present to each visitor.

积累堆叠
Layer Upon Layer Office

项目地点：中国台湾 / Location : Taiwan, China
项目面积：80 平方米 /Area : 80 m²
公司名称：层层室内装修设计有限公司 / Organization Name : Cengceng interior Design

IAI 设计优胜奖

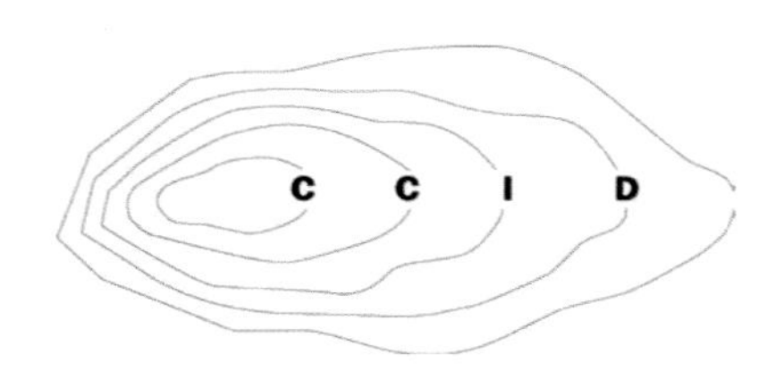

层层设计 Cengceng Design

室内设计是不同元素的搭配组合，透过与屋主的沟通，设计师将灵感、需求、美学、机能等置入空间。
重要核心在于设计的沟通，完工后样貌是屋主的期待，因此 3D 立体图的绘制，就是双方沟通中重要工具，以专业为有限的空间，创造无限的生命力。

Interior design is a combination of different elements. Through communication with the homeowner, the designer puts inspiration, needs, aesthetics, function and so on into the space.
The important core lies in the communication of design. After completion, the appearance is the expectation of the owner. Therefore, the drawing of the 3D perspective is an important tool in the communication between the two parties. With professional space, it creates unlimited vitality.

放大“累积”视觉，合并“堆叠”手法，将设计主轴，以多元纵向的大元素概念，发挥出视觉力度。作品为空间设计之办公室，通过行销端点与设计目标等需求，将灵感、美学、机能、需要等关键养分，一层层地有序置列于群山万壑间，于山峦遍层中循着空间脉络秩序，累积出空间视觉的惊叹雅致之作。

“山”的意象，是层层设计的识别形象，也是对设计热忱的初衷。在层层的办公空间里，醒目的立体切割山形脉络装置，每天提醒着层层团队要以扎实稳健的设计基础，与丰富涵养的专业知识，在空间设计中以巧夺天工的挥洒，创作如山林间变化万千的室内景致。

Zoom in on the "cumulative" vision, merge the "stacking" technique, and design the spindle to give full visual power to the concept of multiple elements in multiple verticals. The work is an office for space design. Through the needs of marketing endpoints and design goals, key nutrients such as inspiration, aesthetics, function, and needs are placed in layers in the mountains. Following the spatial context of the space, accumulating the exquisite and elegant work of space vision.

The image of "mountain" is the identification image of the Ceng Ceng design, and it is also the original intention of design enthusiasm. In the Ceng Cengoffice space, the striking three-dimensional cutting of the mountain-shaped vein device reminds the Ceng Ceng of the team every day to have a solid stability design foundation, and rich professional knowledge, in the space design, sprinkle with wonderful article excelling nature, create the indoor views like ever-changing forests.

拜腾制造基地办公室
Byton Nanjing Office

项目地点：中国南京 / Location :Nanjing , China
项目面积：9890 平方米 /Area : 9890 m^2
公司名称：领筑智造 / Organization Name : inDeco

IAI 设计优胜奖

领筑智造 inDeco

inDeco 领筑智造一站式互联网公共空间设计装修服务机构，专业提供空间设计、装修施工、软装配饰、后期配套等服务。始终以“设计感、高质量、透明化”作为核心理念，拥有极具创意的设计团队、专业化的供应链管理团队及精细、高效的施工团队。

inDeco is The One-stop Solution for Office and Public area Design, Engineering, Technology, Management, Construction and Commissioning. Believing in “Edging design, High quality, Transparency” as the core value. With In-house innovative design team, professional supply-chain team and project management team, inDeco continues to thrive for Efficiency, Delicacy, Transparency.

本项目是来自原有工厂改造，充分结合拜腾设计、科技、未来的核心概念，整体采用工业的设计风格，打造出集办公、休闲、会客及展示的办公综合体；采用拜腾（Byton）品牌以及线条的VI元素完美融入到结构墙体中。考虑其行业个性，将车的行驶轨迹中的迹字本土化处理，采用带有中国特色的水墨字“迹应用到会议室。地面采用抽象化的车迹轨道，进行不规则线性分割，既体现企业性质又提升空间属性。

InDeco’s design fully integrates the core concepts of design, technology, and the future. It adopts an industrial design style and the main colors are back, white and gray to create an office complex that integrates office, leisure, guests and product display. The Byton brand and VI elements are perfectly integrated into the structural wall. Considering industry personality, inDeco has localized tire tracks by transforming them into Chinese calligraphy in the conference room; they are also applied abstractly to the flooring. Irregular linear segmentation, which not only enhances the nature of the enterprise but also enhances the spatial properties

生华
Sublimation

项目地点：中国南京 / Location : Nanjing, China
项目面积：750 平方米 /Area : 750 m²
设 计 师：张兆勇 / Designer : Zhao Zhaoyong

IAI 设计优胜奖

张兆勇 Zhang Zhaoyong

筑维空间设计总监
高级室内建筑师
高级陈设设计师
室内建筑协会会员
南京住宅环境设计专业委员会委员

Director of Zhu Wei Space Design
Senior Interior Architect
Senior Display Designer
Member of Interior Architecture Association
Member of Nanjing Residential Environmental Design Professional Committee

整体空间装饰以运用大理石砖以及橡木等材料来体现空间的整洁、格调的高雅，从材料、工艺、景观、产品等元素与人文要求之间寻求最佳平衡点，将诗意巧妙融入正式的办公环境， 通过材质纹理所表现出的色感延续了自然界的简约韵律，合理地将现代化办公室的元素通过简约大气的方式呈现出来。

The overall space decoration of this case uses marble bricks and oak to reflect the cleanliness and elegance of the space. From the materials, crafts, landscapes, products and other elements to the human requirements, we seek the best balance between poetry and ingenuity. The office environment, the color sensation expressed by the material texture continues the simple rhythm of nature, and the elements of the modern office are reasonably presented in a simple and atmospheric way.

House Space

室内 —— 住宅空间

寒舍
Humble House

项目地点：中国台湾 / Location : Taiwan, China
项目面积：235 平方米 / Area : 235 m²
公司名称：汤尼大卫设计有限公司 / Organization Name : Tony David Interior Design

IAI 设计优胜奖

汤尼大卫 Tony David

汤尼大卫设计公司，创设于 2002 年，是一个专注于质感与创新的设计事务所，业务范围涵盖中台两地，商业空间及住宅均有丰富设计经验，常获颁国内外设计奖项。

The Tony David was established in 2002, it is a design firm which is specializing in texture and innovation of the interior design. The business scope covers both China and Taiwan. However, it has rich design experience in commercial and residential spaces, and it has been awarded domestic and international design awards.

亚洲东方文自古便以寒舍谦称自己的家，是一种对人亲切与虚怀的表示。然而在这单层住宅中，我们希望，家也以这种姿态呈现，开阔的空间格局留给人与人之间直接真切的交流，以质朴醇厚的材质展现壁面质感，并在居家细节中置入带有东方文人的格调的趣味，在此迎接归来之人，以及来访的亲朋好友。东方文人自古便以寒舍谦称自己的家，是一种对人亲切与虚怀的表示。

East Asian Literature has called Humble House its home since ancient times. It is an expression of kindness and humility toward people.However, in this single-story house, we hope that family will also be presented with this attitude. The open spatial layout allows for clear, direct communication between people. The texture of the walls is displayed in simple and mellow materials. There is an oriental scholar style and taste in the details of the home, welcoming those returning home as well as visiting relatives and friends.

湖州精品酒店
Boutique Hotel Huzhou

项目地点：中国浙江 / Location : Zhejiang, China
项目面积：4000 平方米 / Area : 4000 m²
设　计　师: 朗耀 / Designer : Jeorg

IAI 设计优胜奖

朗耀 Jeorg

朗耀师从 Wolf D. Prix 教授（蓝天事务所）并取得了硕士学位，并且朗耀在阿姆斯特丹和上海的 UNStudio 也从事多年。2013 年，朗耀成立了他自己的工作室 LOFE。2013 年，朗耀与 UCLA 的 3M 未来实验室合作，建造了第一栋 3D 打印房屋。这也是他从 2009 年就开始教授的内容。他在同济大学、玻利维亚的天主大学、复旦大学都教过课，做过讲座和演讲嘉宾。

Joerg graduated from the masterclass of Prof. Wolf D. Prix at the university of applied arts in Vienna (Coop Himmelblau) and worked for many years as an associate at UNStudio in Amsterdam and in Shanghai. In 2013 Joerg founded his own office LOFE Studio. Joerg was building the first 3d printed house with the 3M futurelab by UCLA, a format he was teaching since 2009. Other teaching positions include the Tongji University and the Catholic University of Bolivia, Fudan University and various guest crits and lectures.

在传统地方民宿和现代民宿之间创建了一个和谐的平衡点，为每一个在上海周边寻求宁静的人创造了独特的体验。当您从高处远眺时，建筑的布局是看起来像三个站立的被大树遮蔽的农舍。我们的目的和意图是保持这种民宿的外观和感觉以及这些古树，所以，我们重建了这些建筑物以满足现代建筑的设施和设备，同时保持旧建筑以及材料的原始性和魅力。

在山的下方，沿着起伏的景观线，嵌入了现代化设计的客房以及配套设施，包括健身房、日式室外温泉、按摩中心、商务中心、台球和游泳池。酒店形成了一座连接现代和历史、城镇和乡村风格的桥梁，并将其与当地的自然风光融为一体。

Designing a harmonious balance between traditional local farmhouses and modern insertions creates a unique experience for everyone who is seeking a retreat in the close proximity to Shanghai.
The building is arranged that on the top side from which you arrive, the building looks like the three farmhouses, which were standing there and are sheltered by big old trees. It was on purposes and intention to keep this look and feel and the old trees-however the buildings now were reconstructed to cater for modern building equipment whilst keeping the originality and Charms as well as materials from the old buildings.
In the lower side of the Hill following the undulating lines of the landscape two modern designed levels of guestrooms including amenities such as gym, an outside Japanese spa, massage rooms, a business center, a billiard and pool area were inserted. The hotel forms a bridging modern and historic, town and country style and embedding it in harmony to the site.

森林别墅
Forest Villa

项目地点：中国浙江 / Location : Zhejiang, China
项目面积：400 平方米 /Area : 400 m²
设 计 师: 朗耀 / Designer : Jeorg

IAI 设计优胜奖

整个别墅的设计概念是把森林从外面带进屋内并且在房屋的外墙上也得以体现。一个向上螺旋形的“森林带”划分了地域并环绕整个房屋，将房屋的整体布局规划进行垂直和水平的连接并且组成了外墙面。

在3D模型中，螺旋形线条对斜坡地形进行了划分。别墅就位于这条螺旋线上。因此，楼梯从到达层向上蜿蜒而上，在那里房屋被支柱垫高，汽车可以停在悬臂下。楼梯环绕着起居室，一直延伸到房子后面的第二层。在这里，“森林带”连接着用木头包裹的楼梯，延伸到外面并将整个房屋包裹成一个木制的外墙立面。

The concept of the house is to bring the forest from the outside into the house and out again to form the facade. An upward spiraling “forest band” cuts the terrain and wraps around the house – organizing the floorplan as vertical and horizontal connection and forming the facade.

The sloped terrain was cut in the 3d model with spiral. On this spiral the “room boxes” were hanged on. Hence the staircase winds up from the arrival level where the house is elevated on stilts and the cars can park under the cantilever. The staircase wraps the living room and goes up to the second level on the backside of the house. Here the “forest band” the staircase, which is cladded in wood, goes to the outside and wraps around the house as a wooden facade cladding.

玫瑰源别墅
Melrose Hill Villa

项目地点：中国江苏 / Location : Jiangsu, China
项目面积：823 平方米 / Area : 823 m²
公司名称：源点设计 / Organization Name : LOCUS ASSOCIATES

IAI 设计优胜奖

LOCUS°

源点设计 LOCUS°

源点设计坚持创造独特而互融的设计，以表现场所灵魂和地域精神。我们的设计注重通盘考虑，聚焦于对文化历史、当地文脉和周边环境的研究和提炼，运用现代的设计语言，赋予场地元素明确特征和永恒的意义。
源点设计提供一个延续设计思想、设计对话和设计进程的平台，沉淀空间设计中深刻的敏锐性和细微的平衡性。

LOCUS° insists on creating unique and inclusive designs to express the soul of the place and the spirit of the region. Our design focuses on overall consideration, focusing on the study and refinement of cultural history, local context and surrounding environment, using modern design language to give the site elements a clear identity and eternal meaning.
LOCUS° provides a platform for continuation of design ideas, design dialogues and design processes, with deep acumen and subtle balance in sedimentary space design.

在对这个高雅古典的别墅进行设计改造时，我们打破视觉和空间上的框架，让别墅内的每个空间都融合在一起，同时揉合现代设计美学概念，使人享受到深邃、沉稳和闲适的生活态度。

业主是一位当地知名的概念餐厅东主，首层无疑成为一个接待和享用私房菜的派对空间。通过一个狭长的廊厅，连接着餐厅、茶艺室与开放厨房区。业主和他的餐饮团队，在此招待他们的客人。

别墅座落于小山丘上，宽阔草坪和庭园景观倒映在室外的水池中，交织成唯妙的视觉感官。地下室则成为音乐和艺术的休闲空间，展示着业主的文化珍宝与乐器藏品，以现代设计相结合的配搭来创造一个安静沉思的观赏氛围。

At the onset in conceptualizing the interiors of this classical house typology, we wanted very much to break the existing volumetric structure with its interiors. This allows the house better connectivity, both visually and spatially, in order to bring about a modern aesthetic. With the owner being a local fusion restaurateur, the arrival ground floor entry level decidedly becomes the reception and entertainment wing wherein guests are welcomed with a long axial formal gallery linking all spaces in the living, dining, cards rooms and show kitchen spaces. This turns out to be an effective party space for the frequent dinner parties the owner throws.
With the house resting on a hilly terrain, the basement opens outwards towards an expansive lawn and reflecting pool in the lower garden. Setup as a more introvert level for music and visual arts, this level includes the main study, treasures collection synonymous to local Yixing culture of tea pots and cups, as well as the tea and music rooms. It is here that we sort very much to fuse local Chinese culture with contemporary design to bringing about a cohesive dwelling conducive for quiet contemplation and appreciation.

顶层住宅W^2

W^2 Loft

项目地点：中国广州 / Location : Guangzhou, China
项目面积：540 平方米 /Area : 540 m^2
公司名称：北京艾奕空间设计有限公司 /Organization Name : AE Architects Design

IAI 设计优胜奖

顶层住宅W^2位于广州一个历史气息深厚的区域，在一个独特的新小区的最顶部，享受着城市之巅的绝佳景色，5米高的复式结构环绕着引人瞩目的窗户和尽收眼底的城市景观。年轻的新婚业主夫妻特别关注简朴舒适，希望在此创造一个明亮的与家人团聚共享的现代简约住宅。因此，我们选择了天然材料，包括木材、白色大理石和立体绿植墙。为了加强住宅的功能，厨房和餐厅上空增加了一个夹层，成为一个带有天窗的工作室，并连接通往天台的楼梯。鉴于中国南方的气候条件，年中大部分时间都温暖宜人，一个能够开窗，享受阳光、空气与风，产生户外活动的生活空间是必须的。整层主卧室在天台下方，拥有宽敞的浴室，以及分别属于他和她的步入式开阔衣帽间。穿过保证私密性的主卧室大门下楼，孩子和孩子祖父母的卧室完整了这个温暖的家。极简和开放的空间设计语言，契合了年轻新一代的生活方式。

Located on top of the historic district of Guangzhou, occupy the last 3 floors of an apartment tower inside a new and exclusive residential compound, W^2 loft enjoy the best views of the area. The 5m height floor is surrounded by impressive windows and openings to the city.The client wished to take advantage of this situation and create a bright, modern and simple space where gather with the family. This just married young couple paid special attention to the simplicity and comfort of the space.Natural materials like wood, white stones and green walls were chosen. In order to increase the functions of the house, a mezzanine was added over the kitchen and dining area. In this second floor we designed a studio with a top light window and a stair up to the roof top.Given the conditions of the south of China weather, warm for most part of the year, the creation of outdoor activities to enjoy sunlight, open air and wind was a must. Right under this terrace the main bedroom floor enjoys a generous bathroom and spacious walk in closets for him and her.Crossing the private door and down stairs, the kids and grandparent bedrooms complete the house. Minimalistic and open space language fits the lifestyle of this new generations

澳门天钻高级公寓
Macau Natural Diamond Service Flat

项目地点：中国澳门 / Location : Macao, China
项目面积：155 平方米 / Area :155 m²
公司名称：邸设空间设计有限公司 / Organization Name : Decor & Decor Interior Design

IAI 设计优胜奖

邸设设计 Decor & Decor Interior Design (Shanghai)

成立于 2015 年，由从业十几年的国际国内软装主创共同打造的国际设计团队，设计师主要以业内顶尖女性设计师，主打室内硬装、软装设计与产品改造，品牌秉承着对精致生活的追求，打造室内软装设计结合家居产品私人定制设计为一体的软装设计模式。为居住者呈现更舒适、更精致的室内设计同时，融入居住者个人性格与形象气质的私人定制空间。

DiShe（Décor & Décor）Interior DesignCo., Ltd was founded in 2015, the international design team created by the international and domestic soft-wearing designers who have been engaged in the industry for more than ten years, the designers are mainly the top female designers in the industry, focusing on the interior hard-wearing, soft-packing design and product transformation. The brand adheres to the pursuit of exquisite life and builds The soft design of the interior is combined with the custom design of the home product. For the occupants to present a more comfortable, more refined interior design, while incorporating the personal custom of the occupants' personal character and image.

设计不但要满足设计感还需要环保型功能型结合。在环保材料的选择上我们更加注重，如客厅的手织地毯，出自美国画家Fort Street Studio之手，无胶底环保型材料制作，价值17万人民币。我们不但在大件单品上做了思考，更是在小小的摆设上做了精而少的筛选。设计源于生活，服务于生活。只有更多关注生活，才能带给客户良好的体验。

In this house, we have paid more attention to the selection of environmental protection materials. For example, the handmade carpet in the living room, out of hands by the American painter, Fort Street Studio, is made from environmental protection materials without any rubber sole. The value of this carpet is worth 170, 000 RMB. As you can see that, we have not only pondered over the disposal of large-sized objects but also make a dedicated screening for those small furnishings. Design comes from life and will also service for life. Only if we focus more on details of life, will we be able to bring the better experience to our customers.

林夕
Forest Twilight

项目地点：中国南京 / Location : Nanjing, China
项目面积：450 平方米 / Area : 450 m²
设 计 师：顾伟伟 / Designer : Tony Gu

IAI 设计优胜奖

顾伟伟　Tony Gu

南京龙瑞装饰设计工程有限公司总监级设计师
中国建筑装饰协会高级室内建筑师
2018 星河 · 第三空间 "IDS 大奖" 别墅空间设计奖
2018 第六届中国装饰设计奖 银奖
2018 "IAI 设计奖" 设计优胜奖
2018 法国双面神 "GPDP AWARD" 国际设计大奖提名奖

Nanjing Longrui Decoration Design Engineering Co., Ltd.
Director-level designer
Senior Interior Architect of China Building Decoration Association
2018 "IDS Awards"—Villa Space Design Award
2018 The 6th China Decorative Design Award—Silver Award
2018 "IAI DESIGN AWARD"—Design Excellence Award
2018 "GPDP AWARD"—Nomination Award

简约的空间可以让我们获得相对轻松、舒适的生活环境，简洁和实用是简约风格的基本特点，它以追求室内空间的简略、摒弃不必要的浮华而大行其道。与传统风格相比，简约设计所呈现的是剔除一切繁琐的装修设计元素，用最直白的装饰语言体现空间和家具所营造的氛围，进而赋予空间以个性和宁静。

The simple space allows us to get a relatively relaxed and comfortable living environment. The simplicity and practicality are the basic characteristics of the simple style. It is popular in pursuit of the simplicity of the interior space and the elimination of unnecessary flash. Compared with the traditional style, the minimalist design presents all the cumbersome decoration design elements, and uses the most straightforward decorative language to reflect the atmosphere created by the space and furniture, thus giving the space personality and tranquility.

燕西华府 42# 合院别墅
Yanxi Huafu 42# Courtyard Villa

项目地点：中国北京 / Location : Beijing, China
项目面积：750 平方米 / Area : 750 m²
设 计 师：黄婷婷、杨俊 / Designer : Huang Tingting, Yang Jun

IAI 设计优胜奖

黄婷婷 Huang Tingting
杨俊 Yang Jun

天鼓设计创始人及设计总监，致力于室内设计行业二十年，具有建筑及室内结构学术背景，擅长从建筑出发，结合景观，做到建筑与室内一体化设计。追求完美精品，深得地产甲方肯定，并获得多项室内设计大奖。

As co-founder and chief designer of Shanghai Tian Gu Interior Decoration Design Co. Ltd, Mr. Yang has dedicated himself to interior design for 2 decades. Being an expert in building industry and structural engineering, he is skilled in extraordinary integration of architecture and interior design. Pursuing perfection has always been his goal. His works are constantly appreciated by a great many real estate developers. He has won a number of interior design awards.

古代医典《灵枢》中写到："春生、夏长、秋收、冬藏，是气之常也，人亦应之。"天人合一，顺应四时，二十四节气以及对应节气的吃穿住行都是学问。我们将这些理念强调在住宅的每一个空间里，并通过历代拥有美好寓意和诗意的物象来表达最适宜中国人居住的人文居所。设计师从空间设计到软装配饰都着力强调阳光，强调宜居，强调温情，强调人与自然的和谐共生。更希望通过设计，让更多人和年轻一代对中国文化的传承产生共鸣。

In one of the ancient Chinese medical books, four seasons and the 24 solar terms are defined according to the position of the sun on the ecliptic. From the perspective of the Chinese, , what is most important in all the important decisions in our lives is to do the right thing. Second, and only slightly behind the first, is to do the right thing at the right time. We apply these theories to every corner of the house, representing the common wish of Chinese people for a better living and creating blueprints for harmonious life styles.
The designers have taken daylight performance into account and created buildings that deliver a much better occupant experience. What they have succeeded in is that they have raised our awareness and ability to find meaning in the life around us and reminded us of the importance of cultural heritage.

建筑师之家
Home of architect

项目地点：中国上海 / Location : Shanghai, China
项目面积：380 平方米 / Area : 380 m²
设 计 师：赖仕锦 / Designer : Lai Shijin

赖仕锦 Lai Shijin

先后在多家国外设计机构上海分公司担任专案设计师、设计总监。室内设计项目遍及各省市及东南亚等国家，涉及多个领域的室内设计，擅长酒店、办公室、商业及住宅设计。

He has worked as a project designer and design director for a number of foreign design agencies in Shanghai. Interior design projects in various provinces and cities and Southeast Asia, involving interior design in a variety of areas, specializing in hotel, office, commercial and residential design.

本案是一栋"3+1"户型的别墅设计，男主人是一名来自中国香港的建筑师，职业的关系让他对空间细节和整体布局有着高于常人的要求，简洁和通透是他理想的居住环境的表达。因此，在满足功能需求的基础上，设计师始终遵循"less is more"的黄金法则，让设计回归纯粹的本质，从而营造出一个自然舒适的空间氛围。

由于建筑结构本身的局限性，为了能充分发挥出大空间的优势，设计师通过拆掉一些墙体，并结合玻璃隔断的层次分割，最大程度上利用采光和空间结构，去塑造一个通透的空间。

This case is a "3+1" villa design. The male host is an architect from Hong Kong. He has higher requirements on space details and overall layout. The concise and transparent is the ideal living environment for him. Therefore, on the basis of satisfying functional requirements, the designer always follow the golden rule of "less is more" to return the design to the pure nature, so as to create a natural and comfortable space atmosphere.

Due to the limitations of the building structure itself, in order to give full play to the advantages of large space, the designer removed some walls. Combining the level of partition of glass partition, he made use of daylighting and space structure to shape a transparent space.

爱很简单
Simple Love

项目地点：中国河北 / Location : Hebei, China
项目面积：750 平方米 /Area :750 m²
设 计 师：李明锋 / Designer : Li Mingfeng

IAI 设计优胜奖

李明锋 Li Mingfeng

美国 CRAZY DESIGN（简称 CDN）设计师事务所联合创始人
香港弘盛嘉合设计事务所设计总监
弘泽拓展装饰工程有限公司董事长
至简未来家居体验中心设计顾问
国际建筑装饰室内设计协会高级室内设计师

American CRAZY Design(CDN) Co-fonder of design firm.
Hong Kong Hong Sheng Jia He Design Firm Design Director.
Hong Ze Tuo Zhan decorative Engineering Co.,Ltd. Chairman.
Zhi jian Future Home Experience Center Design Consultant.
Senior Interior Designer of International Construvtion Decoration interior Design Association.

本案命名“爱很简单”，是夫妻二人内心的写照，人简单，爱简单，房简单。四层别墅，总面积750平米，经过大胆的重构，化繁为简，整个空间利用率达到最大化，空间宽敞大气，整体风格简单纯净。软装采用“禅”的理念，让空间有在简单的氛围下，给人无限的遐想空间。

The case named love is very simple. It is the portrayal of the two people of the husband and wife. It is simple people, simple love, simple house, four storey villa and 750 square meters in total area. After a bold reconstruction, it is simplified, the utilization rate of the whole space is maximized, the space is spacious and the whole style is simple and pure. The concept of "Zen" is used in soft clothes, so that space can give people unlimited reverie in a simple atmosphere.

三角之家
Delta Home

项目地点：中国台湾 / Location : Taiwan, China
项目面积：50 平方米 / Area : 50 m²
设 计 师:李培媺、赖昱承 / Designer : Li Peiwei, Lai Yucheng

IAI 设计优胜奖

公共空间、客厅、餐厅、厨房的空间，通过借用空间进行视觉设计，增强了“家庭氛围”的流动感，开放的系统架构也使得餐厅和厨房不仅可以成为供餐的地方，而且还可以成为家庭聚会的休闲空间，高冷的颜色被选为空间的基调，棕色、米色、黑色和白色的不同色调形成了互补而层叠的外观，整体空间结合了成熟、极简主义和一丝活力，阴与阳的柔和混合。

The arragement of public space, living room, dining room, kitchen, designs by borrowing space to make visual penetration, strengthening a sense of " home-atmosphere " flows. Open system architecture also allows the restaurant and kitchen not only become a place to provide, but also be a leisure space with family gathering.Cold colors are selected to be the base tones of the space, varying shades of brown, harmonious beige, black, and white form a complementary yet layered look.The integral space combines both mature, minimalism and a hint of vibrancy, a gentle mixture of Yin and Yang.

李培媺 Li Peiwei

空间设计都蕴含着丰富的经验，身临其境感受周围的环境，通过大量手的温度去感受生活，互相运作达成我们的愿景。

Space design all contain a wealth of experience, immersive feel the surrounding environment, through a lot of hands to feel the temperature of life, functioning each other achieve our vision.

赖昱承 Lai Yucheng

我喜欢旅行，喜欢在大自然散步吹风放空自己的身心，灌溉转换的心思在设计里，用真挚的人生如洗练作品，成就工作者的最大价值。

I like to travel, like to walk in nature to blow out their body and mind, irrigation conversion of the mind in the design, with sincere life such as washing works, the achievement of the greatest value of workers.

KELLY HOPPEN HOME

域
The Domain

项目地点：中国台湾 / Location : Taiwan, China
项目面积：80 平方米 / Area : 80 m²
设 计 师：李培媺、林轩慧 / Designer : Li Peiwei， Lin Xuanhui

IAI 设计优胜奖

李培媺 Li Peiwei

空间设计都蕴含着丰富的经验，身临其境感受周围的环境，通过大量手的温度去感受生活，互相运作达成我们的愿景。

Space design all contain a wealth of experience, immersive feel the surrounding environment, through a lot of hands to feel the temperature of life, functioning each other achieve our vision.

林轩慧 Lin Xuanhui

生活太复杂，我想专注地体悟环境，从习以为常的事物中，挖掘不同的美妙之处，这就是空间设计及生命的本质。

Life is too complicated, I would like to focus on understanding the environment, from things for granted, the mining of different beauty of this is the essence of life and space design.

公共空间的多余隔墙已被移除，形成一个开阔的大型居住区，通过设计与传统不同的平面布局，使自然光线与家庭互动更多，线条的简洁组合是这个项目的主要风格，可以带来清新的感觉，通过在一个大的地方充分利用白色来创造重要性、清洁、和平和耐用性，我们更关注意境而不是简单地进行正式的休息，因为我们喜欢纯洁的极简主义。

An excess partition has been removed in the public space, to create a large and open living area.By design different plan layout from tradition, to bring the natural light in and more interaction with family.The concise composition of lines is the main style of this project, which can bring out the refreshing feelings. We are able to create significance, cleanness, peace and durability by making the best of white in a large place. We focus more on artistic conception rather than simply making a formal break. Because we love minimalism for its pureness.

五十度灰
Shades of Grey

项目地点：中国台湾 / Location : Taiwan, China
项目面积：750 平方米 /Area : 750 m²
设 计 师:李培微、汤伊芳 / Designer : Li Peiwei , Tang Yifang

IAI 设计优胜奖

李培微　Li Peiwei

空间设计都蕴含着丰富的经验，身临其境感受周围的环境，通过大量手的温度去感受生活，互相运作达成我们的愿景。

Space design all contain a wealth of experience, immersive feel the surrounding environment, through a lot of hands to feel the temperature of life, functioning each other achieve our vision.

汤伊芳　Tang Yifang

审视我内心的声音，观察对事物的敏锐，我透过对色彩的优势，用真诚的态度传达给你们，这就是我对于设计的初衷。

Look at my inner voice, observant of things, I through my advantages of color, to convey to you with a sincere attitude, this is my original intention for design.

室内设计涉及与自然环境的相互作用关系，其中太阳、空气和水是非常重要的。对于那些使用空间的人来说，生活方式和空间视觉是每天的一部分，在设计时，我特别强调细节的作用和精心挑选的材料。我相信最天然的材料，如木材、石材纹理和玻璃的力量，应用经典的灰色调色板，设计师希望将现代风格与低饱和度色彩相结合。此外，通过将大面积的空间留下白色，它会在平面之间创造出微妙有序的感觉，它们的独特性重新定义了整个空间。

The interior design concerns the interaction relationship with the natural environment, of which the sun, air and water are of great importance. For those using spaces, the life style and space visual are part of every day.I believe in the power of the most natural materials, such as stone, natural wood, stone art texture and glass.Applying the classic gray-tone palette, the designer hopes to combine the modern style with low saturation colors. Also by leaving a large area of the space white, it creates a subtle feeling of order between the planes. Their unique character defines the whole space.

时代
Era

项目地点：中国浙江 / Location : Zhejiang, China
项目面积：450 平方米 /Area : 450 m²
设计师：尹晓敏 、沈丹 / Designer :Yin Xiaomin， Shen Dan

IAI 设计优胜奖

尹晓敏 Yin Xiaomin
沈丹 Shen Dan

浙江后朴设计合伙创始人。设计不是一种技能，而是捕捉事物本质的感觉能力和洞察能力，并执着于将设计演绎到极致。擅长室内设计，环境设计，设计了包括了精致私宅、精品商业空间、办公空间、高档别墅等设计项目。

Founder of Houpu Design Partnership in Zhejiang Province
Design is not a skill, but the ability to grasp the essence of things, sense and insight, and persevere in deducing the design to the extreme, creating wonderful design works repeatedly. Being specializing in interior design and environmental design. havedesigned many design projects including delicate private house, fine commercial space, office space and high-grade villas.

本案的设计中充分地结合木材与瓷砖，主次之分恰到好处，每个空间里都能完美地展现出各种材质的特色，电视背景、沙发背景、过道背景、每个瓷砖的背景旁都是木材默默地衬托，就好像眼前的景象就是男主人宠爱和包容妻子的场景一样，是多么地温馨和幸福，我想这就是生活与设计最好的结合。

In the design of this case, wood and ceramic tiles are fully combined, and the primary and secondary points are just right. Each space can perfectly show the characteristics of various materials, such as TV background, sofa background, aisle background, and the background of each ceramic tile is silently set off by wood, just as the present scene is the scene where the male owner dotes on and embraces his wife, how warm and happy , it is I think this is the best combination of life and design。

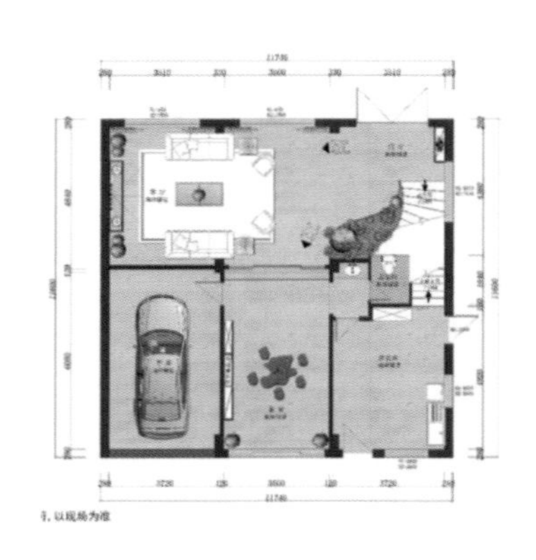

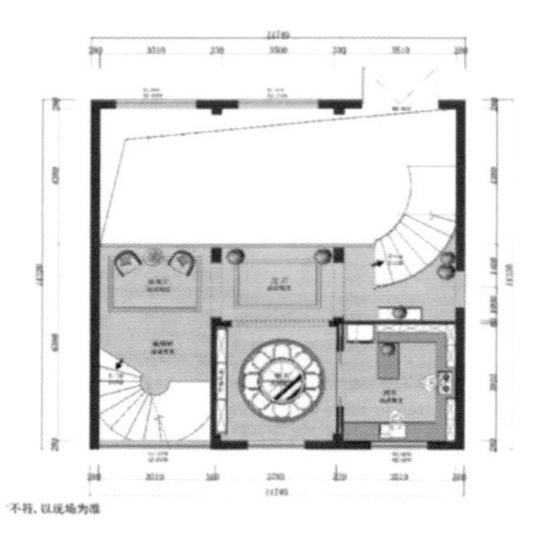

引光
Led Light

项目地点：中国台湾 / Location : Taiwan, China
项目面积：149 平方米 /Area : 149 m²
公司名称：树屋室内装修设计有限公司 / Organization Name : Tree House Design

IAI 设计优胜奖

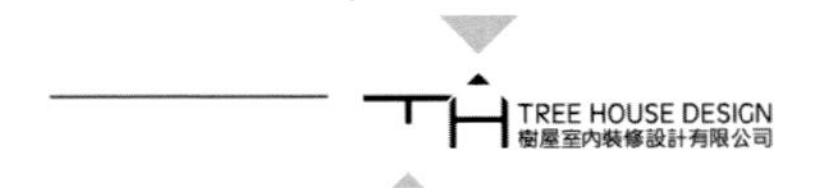

树屋室内设计
Tree House Design

树屋设计是一支以梦想与执行力所组成的设计团队，抱持着对美学的绝对坚持、独特的创意发想，针对居者的需求与个人生活习性，展现细腻的关怀与深度的美学是创办人坚持的理念，辅以精确扎实的施工品质，为业主打造一件件隽永而能温暖人心的作品。

Tree House Design is a team of young designers, drawing creative thoughts from our aesthetic point of view with fine execution. We believe our design takes consideration of the needs of our residents and their styles of living, which reflects through the details of precise construction. By combining what our clients want and our aesthetic concept, our timeless state-of-the-art design will always be the home to return to.

整栋建物引光入室点亮生活起居的动线，空气流转畅通无阻，从天井洒落的光线折射在每个角落晕染出不同的空间表情，为求不破坏主体建物的优良条件。本案仅采取半留白低度设计，墙面留白像画布般捕捉光和影，入口天花板以十字黑色灯沟一横一竖直接剖出主动线，利落简快划分出客餐厅区域，亦让空间绵延蕴籍。餐厅的多功能吧台及餐桌，使用系统柜搭配建商原有厨具配色，可拉伸旋转的餐吊灯，将圆形语汇点缀洗练的细节之中。天花板上的空调维修孔及冷气出风口，搭配灯具做一体性沟缝的设计，突出空间的细腻质感，设计看似简洁却不显单调，些许的黑与白、温润的木皮相互交织激荡，让空间满盈纯粹之美。

The whole building enlightens the movement of living with light. The airflow in the house is fluent, and the light sheds through and scatters in every corner to color different spacial expressions. To upkeep the good condition of the main building, this case takes on a semi white low level design. The white wall captures the light and shadow just like a canvas. The ceiling at the entrance is divided with a cross black light to sharply divide the living room and dining room, so as to extend the sense of space to the multifunctional bar and dining table. The dining room has a matching colour and cupboards. The hanging lights in the living room are turned on with pull strings and decorated in delicate details with its round shape, the air-conditioning vent on the ceiling is combined in unity with this circular theme. The design looks simple but not simplistic. The white and black, and a warm wooden skin weave together are bring this room to life with the elegance of purity.

湖州绿城御园49号别墅
No.49 Greentown Royal Garden Villa Huzhou

项目地点 : 中国湖州 / Location : Huzhou, China
项目面积 : 1200 平方米 / Area :1200 m²
设 计 师 : 王建 / Designer : Wang Jian

IAI 设计优胜奖

王建
Wang Jian

国家注册室内设计师
中国高级室内建筑师
中国建筑协会室内设计分会会员
中国建筑装饰协会会员
中国室内装饰协会会员
全国杰出中青年室内建筑师
全国住宅装饰装修行业优秀设计师

National registered interior designer
Chinese senior interior architect
Member of China Architecture Association Interior Design Branch
Member of China Building Decoration Association
Member of China Interior Decoration Association
National Outstanding Young and Middle-aged Interior Architect
Excellent designer in the national residential decoration industry

延续了建筑外观的风格，以古典欧陆华贵的法式风格为基调，并且还融入了现代轻奢风，采用去繁就简的手法，给人以浪漫迷人的基调，仿佛普罗旺斯的阳光，洒满你的家。本案的几大设计要素：1.布局上突出轴线的对称、恢宏的气势、豪华舒适的居住空间。2.贵族兼时尚风格，高贵典雅简洁舒畅。3.细节处理上运用了法式廊柱、线条，制作工艺精细考究。4.现代设计手法点缀在自然中，崇尚冲突之美。

Our interior design is also a style that continues the appearance of the building. It is based on the classical European style and the French style. It also incorporates the modern light luxury style. It uses a simple solution way to give a romantic and charming tone, as if The sunshine of Provence is full of your home. The design elements of this case are as follows: 1. The layout is prominent in the symmetry of the axis, the magnificent momentum, and the luxurious and comfortable living space. 2. nobility and fashion style, noble and elegant, simple and comfortable. 3. the details of the treatment of the use of French colonnades, lines, fine and exquisite craftsmanship. 4. Modern design techniques are embellished in nature, advocating the beauty of conflict.

1° 灰
Ethereal Grey

项目地点：中国江苏 / Location : Jiangsu, China
项目面积：150 平方米 / Area : 150 m²
设 计 师:徐振 / Designer : Xu Zhen

IAI 设计优胜奖

徐振 Xu Zhen

集室内设计师、平面设计师、全屋解决方案职人于一身的设计师，徐振（澜本创始人）擅长汲取生活中的精髓，通过极具生活美学的诠释方式将其展现。过去几年设计生涯中，带领公司荣获多个本地及国外设计奖项。他的设计风格鲜明，深受新老客户推崇，为公司年轻设计师及助理带来启发。

Xu Zhen (founder of L.AB-studio) is a design organization that integrates interior designers, graphic designers and whole-house solution professionals. Xu Zhen is good at drawing the essence of life and presenting it through the interpretation of life aesthetics. Over the past few years, the company has won many local and foreign design awards. His design style is distinct and highly praised by both old and new customers, which inspires young designers and assistants of the company.

《1°灰》，这是本案设计师给予作品的名称，介于灰白间的轻柔，似白非灰，灰中显白；难以形容的色彩观感，犹如一滴墨汁沉入一池清水，见而不显，显而又见。置身其中，让内心舒适、安详与平静，儒雅的气质尽显其中，在当下社会，犹如一股清流，直击现代人内心对自然的向往与对曾经过往生活的缅怀，以包容的心态去面对当下世俗间的嘈杂与纷争。

在家具软装的布局上，简洁明朗的线条以木为载体，勾勒出干净的画面，在1°灰的背景下尤其突显。理念上以轻日风格为核心，围绕“断舍离”展开，体现出极简家居生活内涵。业主的生活方式与其家居风格也极为相似，入住半年，也只是简单的收拾下便开始了本次的实景拍摄。这里记录的并非新居起始的原点，而是真实可感知到温度的寻常生活。

"1 degree gray" is the name given to the work by the designer of this case. It is gentle between gray and white, like white but not gray, and white in gray. The indescribable sense of color is like a drop of ink sinking into a pool of clear water, visible but not obvious, obvious also seeable. Place oneself in it, let inner comfort, serenity and calm, the refined temperament fully manifest among them, in the current society, like a clear stream, directly attacking the modern people's inner yearning for nature and the memory of past life, with an inclusive attitude to face the current mundane noise and disputes.
In the layout of soft furniture, the simple and clear lines take wood as the carrier, outlining a clean picture, especially in the background of 1 degree gray. Conceptually, light-day style as the core, around the "break-up and departure" launched, reflecting the minimalist connotation of home life. The owner's lifestyle is very similar to his home style. He stayed for half a year and started the scene shooting simply after tidying up. The record here is not the origin of the new residence, but the ordinary life with real temperature perception.

合一墅
HY.HOUSE

项目地点：中国福建 / Location：Fujian, China
项目面积：1200 平方米 / Area：1200 m²
设 计 师：游小华 、 陈东升 / Designer：You Xiaohua， Chen Dongsheng

IAI 设计优胜奖

游小华、陈东升
You Xiaohua，Chen Dong-sheng

设计总监游小华和陈东升坚持认为“在设计中做好一生”。设计不在于风格。首先，设计是为生活而设计的。更重要的是，将生活与空间美学完美结合。

到目前为止，该公司已经赢得了许多国内室内设计奖项：Idea-Tops、Nest 奖、金堂奖、亚太设计奖。

Design directors You Xiaohua and Chen Dongsheng insisted that “doing a good life in design” . Design is not about style. First of all, design is designed for life. More importantly, the perfect combination of life and space aesthetics.

So far, the company has won many domestic interior design awards: Idea-Tops, Nest Award, Golden Hall Award, Asia Pacific Design Award.

设计起始我们考虑的是：如何让生活和美学和谐统一？如何凝聚空间的形和神（设计灵魂）？如何把美的生活态度与空间完全融合？

幸运的是，《合一墅》的主人HY先生和妻子对美好生活的向往深深地打动了我们，并且用实际行动全力配合项目落地。美好的生活态度需要用实践来证明，需要不断地提升自己的修为，最终回归自己的内心。整套设计主要以白色、灰黑色、线条与块面，独立又相近，相融统一于整体的空间结构与气质，也将现代设计的简约之美引至更深的境地，干净的颜色配以暖白的灯光使整个空间感觉写意并且踏实。

At the beginning of design, we considered: How to unify life and aesthetics harmoniously?
How to agglomerate the form and spirit of space (design soul)?How to integrate the life attitude of beauty and space completely?

Fortunately, we are deeply moved by Mr. and MS. HY's yearning for wonderful life, who are the master of "HY · House" and fully cooperate with the project implementation by practical action. A good life attitude needs to be proved by practice, and needs to constantly upgrade cultivation and finally return to hearts. The overall design is mainly composed of white, dark gray, lines and block surfaces, which is both independent and close; it is integrated into the holistic spatial structure and temperament and also brings the simple beauty of modern design to a deeper sense; the matching of clean color with warm white light makes the whole space freehand and dependable.

白麓
White Foothills

项目地点：中国成都 / Location : Chengdu, China
项目面积：1000 平方米 / Area : 1000 m²
设 计 师：余颢凌 / Designer : Yu Haoling

IAI 设计优胜奖

余颢凌 Yu Haoling

从事室内设计行业 20 年
生活美学践行者、高级室内建筑师
四川尚舍生活设计有限公司创始人
尚舍生活设计总监
Studio.y 余颢凌事务所创始人、设计总监
凌尚舍陈设艺术馆创始人、创意总监
中国室内装饰协会陈设艺术专业委员会副秘书长
BCSD 柏城上建筑设计咨询（上海）有限公司合伙人

Engaged in interior design industry for 20 years
He is a practitioner of life aesthetics and a senior interior architect
Founder of sichuan Shang She home design co., LTD
Shang She design director
Studio. Y Yu Haoling firm's founder, design director
Founder and creative director of ling shang she display art gallery
Deputy secretary general of the display art committee of China interior decoration association
Partnerof BCSD baichengshang architectural design consulting (Shanghai) co., LTD

本案位于成都国际城南天府新区的麓山国际社区，由于该社区的定位是以传统美式生活别墅为主，其空间结构与国人的现代生活方式存在较大的差别，为了解决这一问题，设计师对空间进行了创造性的改造，将一个原本600平米的精装美式别墅，变成了如今1000平的现代主义别墅。以白色为主调，明净通透，铺陈开一曲唯美流动的艺术乐章。

This project is located in the LUSEHILLS international community in Tianfu new district south of Chengdu international city. Due to the positioning of the community is mainly focusing on the traditional American villa, the spatial structure of the community is quite different from the modern life style of the Chinese people. In order to solve this kind of problem, the designer made creative changes to the space with turning and transforming an originally 600 square meter hardcover American villa into a modern villa of 1000 square meters. Being choosing the white as dominant hue, it is clear and transparent with making a beautiful and flowing artistic movement.

鎏金岁月
Precious Time of The Past

项目地点：中国南京 / Location：Nanjing, China
项目面积：1350 平方米 / Area：1350 m²
设 计 师：张兆勇 / Designer：Zhang Zhaoyong

IAI 设计优胜奖

张兆勇　Zhang Zhaoyong

筑维空间设计总监
高级室内建筑师
高级陈设设计师
室内建筑协会会员
南京住宅环境设计专业委员会委员

Director of Zhu Wei Space Design
Senior Interior Architect
Senior Display Designer
Member of Interior Architecture Association
Member of Nanjing Residential Environmental Design Professional Committee

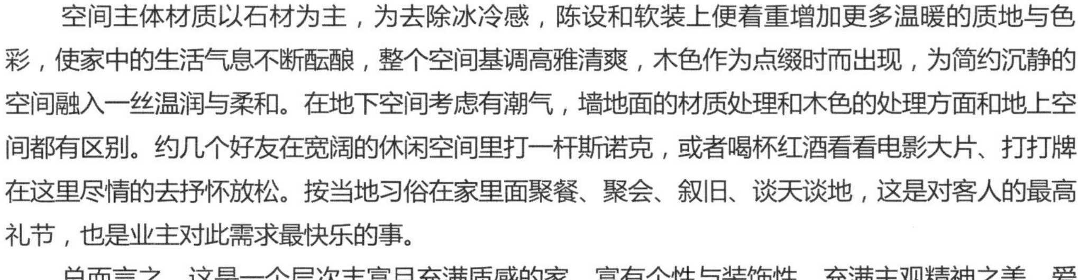

空间主体材质以石材为主，为去除冰冷感，陈设和软装上便着重增加更多温暖的质地与色彩，使家中的生活气息不断酝酿，整个空间基调高雅清爽，木色作为点缀时而出现，为简约沉静的空间融入一丝温润与柔和。在地下空间考虑有潮气，墙地面的材质处理和木色的处理方面和地上空间都有区别。约几个好友在宽阔的休闲空间里打一杆斯诺克，或者喝杯红酒看看电影大片、打打牌在这里尽情的去抒怀放松。按当地习俗在家里面聚餐、聚会、叙旧、谈天谈地，这是对客人的最高礼节，也是业主对此需求最快乐的事。

总而言之，这是一个层次丰富且充满质感的家，富有个性与装饰性，充满主观精神之美。爱生活，多几分亲情，当累了的时候，偶尔回归故里，看看远山溪流，呼吸一下清新的空气，一种舒适与自在的放松状态。

在这个家中，享受生活是一种幸福！

The main material of the space is stone. In order to remove the cold feeling, more warm texture and color will be added to the furnishings and soft clothes, which will make the life atmosphere of the home brew continuously. The whole space is elegant and refreshing, and the wood color will appear as ornament, so as to blend the simple and quiet space with a touch of gentleness and softness. Considering moisture in underground space, the material treatment of wall and floor and the treatment of wood color are different from the space above ground. About a few friends play snooker in the broad leisure space or have a glass of red wine to watch the movie blockbuster, play cards here to express their hearts and relax. According to local customs, it is the highest etiquette for guests and the happiest thing for owners to have dinner, gathering, recounting the past and talking about the world at home.

In a word, it is a home with rich levels and full of texture, full of personality and decoration, and full of subjective spiritual beauty. Love life a little more family love, when tired, occasionally return home, see the distant mountains and streams breathe fresh air, a relaxed state - comfort and freedom.

In this family, enjoying life is a kind of happiness!

黄宅
Huang Residence

项目地点：中国广州 / Location : Guangzhou, China
项目面积：128 平方米 /Area : 128 m²
设 计 师：郑小馆 / Designer :Lauren Cheng

IAI 设计优胜奖

郑小馆 Lauren Cheng

获广州美术学院本科学士学位，十年设计生涯，专注设计工作之余亦曾担任各院校客座讲师。
2016年创立广州深点室内设计有限公司，其创立的设计品牌深点设计，是对自己作品的诠释。

Bachelor's degree of Guangzhou Academy of Fine Arts, ten years of design career, in addition to concentrating on design work, he has also served as guest lecturer in various colleges and universities.
Establish Guangzhou Shen Dian interior design CO.Ltd in 2016, its design brand is the interpretation of its own works.

人间存一角，聊放侧枝花。

欣然亦自得，不共赤城霞。

在这个128平米的三口之家里，舍弃欲望，清空俗事。在龙湖山庄旁，在清幽环境中，投入空灵温暖的臂弯。《卧虎藏龙》里，李慕白曾说过：你握紧拳头，手里什么都没有。你松开十指，却能拥有整个世界。这家人松开十指，舍弃良多，却拥有更简单的幸福。这便是设计师的初衷吧！

Forget desire and clear mundane affairs in the 128m2 house. Around the Longhu Villa and the quiet environment, bump into the hug of warmth. As what Li Mubai in Crouching Tiger Hidden Dragon said, when you clench your fist, you get nothing. When you open your fist, you own the world. This family opened their fists, abandoned so many, and owned more simple happiness. This may be the designer' soriginal intention!

简约质
Style of Simplicity

项目地点：中国台湾 / Location : Taiwan, China
项目面积：181 平方米 / Area : 181 m²
公司名称：山澄有限公司 / Organization Name : Shan Cheng Interior Design

IAI 设计优胜奖

SHAN CHEN
INTERIOR

山澄有限公司
Shan Chen Interior

秉持原则：材与质，取于自然，对环境、居住者，不造成伤害；人与居，注入连结，有故事，有共识，造就一个家。
兼附实用、耐用，符合居者生活习惯，让生活空间有意义；注重美学，将艺术美学注入生活，体现美的视觉飨宴。

Principles Of Adherence : materials Are Obtained From Nature , Which Does Not Cause Harm To The Environment And The Resident. Build The Connection Between The House And The Resident With Stories And Harmonies.
Make The House Practical, Durable And Conformable To Meet The Resident' s Living Habit And Create A Meaningful Living Space. Focus On Aesthetics And Add It To Our Lives With The Visual Feast Of Beauty.

大空间采用一贯性的开放式格局。开放式厨房与餐厅衔接一气，以灰色系调性使空间散发出沉稳质感，料理区紧邻用餐区，拉近家人间的互动。而开放式书房紧邻于沙发区后方，供屋主在办公阅读之余，亦能与家人互动。

整空间公私领域多以大面白色系为基底，使空间感更为开阔明亮。卧房区更强调白色基底质感，以打造放松舒适的睡眠氛围为主轴。主卧以简约的整墙面柜体消化收纳需求，梳妆区域也以白色推拉门板收起视觉杂乱感，简单的开阖收纳规划，即让空间视觉更显清新优雅。

Persistently showing the large open space, the adjoining kitchen and dining room painted in gray colors diffuse mild quality. For the dining table adjacent to the cooking area, the dining atmosphere is pleasant. The open study room, behind the living room couch, brings closer the family connections while offering a cozy working space in the house concurrently.
The large-scale white paints broaden the visual scale in the house. Furthermore, the bedrooms embellished in whites bring forth the complacent sleep environment. In the master bedroom, the massive wall closets and the dressing table furnished with the white sliding door give rise to a clean and orderly personal space as well.

并非世外，只是桃源

Not beyond the Noisy World, Merely A Land of Peace

项目地点：中国上海 / Location : Shanghai, China
项目面积：200 平方米 / Area : 200 m²
设 计 师： 邹洪博 / Designer : Born Zou

IAI 设计优胜奖

邹洪博 Born Zou

曾为多个高端住宅、酒店项目担纲主创设计，拥有十多年的室内设计经验。作品提倡人与空间的互动，追求创新，强调品味、舒适的空间环境。他希望可以通过多年积累的设计经验，为国内的客户提供国际化的设计理念及优质的设计服务，给室内设计行业带入一股新的能量。

Born Zou has worked as a master designer for several high-end residential and hotel projects and has more than 10 years of interior design experience. The work promotes the interaction between people and space, pursues innovation, and emphasizes the taste and comfort of the space environment. He hopes that through years of accumulated design experience, he will provide domestic customers with international design concepts and high-quality design services, bringing a new energy to the interior design industry.

繁华城貌之中的灼灼意境，“只”是人家，温馨美满，引入原木生活、绿色宜居理念。结合当代文人气质，用丰沛的艺术语言与感染力回归内在的价值观与文化诉求，写意韵味盎然、自然意趣的品质生活。让室内演绎了一场与大自然的亲密接触。

“木”本承载，很舒适，有温度。在时光里，无与伦比的美丽。本身自然造诣与“院子”相融，虚与实、远与近之间，空间内瞬间生动，有价值，有魅力。每个细节都值得推敲，兼具功能，设计统一。推开入户门，舒适简约的气息迎面而来，在客厅空间里，米白的沙发，柔软是家的主旋律，营造了令人着迷的归属感。

The burning artistic conception in the prosperous city appearance, "only" is the family, warm and happy, introduce the concept of log life, green and livable. Combining with the temperament of contemporary literati, the rich artistic language and appeal are used to return to the inner values and cultural appeals, and the freehand life is full of charm and natural interest. Let indoor deduce a close contact with nature.

"wood" this load, very comfortable, temperature. In time, unparalleled beauty. Its own natural attainments and "courtyard" are integrated, virtual and real, far and near, space within a moment vivid, valuable, attractive. Every detail deserves scrutiny, function and design unification.

推开入户门，舒适简约的气息迎面而来，在客厅空间里，米白的沙发，柔软是家的主旋律，营造了令人着迷的归属感。 家人攀谈的时光当然留与吧台，一个个美好的故事在这里停留，是家的味道。一个串联的活动区间，天马星空的幻想也能在生活里实现。

Push open enter a door, the breath of comfortable contracted face to come, in sitting room space, the sofa of rice white, softness is the main melody of the home, built the sense of belonging that makes a person infatuate. Of course, the time for family chatting is reserved with the bar, where a good story stays, which is the taste of home.A series of activities, tianma starry sky fantasy can also be realized in life.

海天居
The Sky Residence

项目地点：中国澳门 / Location : Macao, China
项目面积：4310 平方米 /Area : 4310 m²
设计师 / Designer : Eric Tam

IAI 设计优胜奖

"马"元素在内部的许多空间展示。根据中国文化，马的象征有很多意义和历史，它代表着力量、成功、高贵和运气。在生活区域，焦点是墙上的红色女性画，在整个空间中结合了大量的花朵，以增强女性的感觉和层层空间。为了达到设计目标，设计风格、色彩和装饰大多与20世纪30年代的上海时代有关。设计师还在复古的上海主题空间中建立了最新的智能家居系统，使其成为现代技术。

The "Horse" element has continued showing at many spaces of the interior. According to Chinese culture, the symbol of the horse has a great deal of meaning and history, it represents power, success, nobility and luck. In the living area, the focal point is the red female painting on the wall, combining a great amount of flowers all over the space to enhance the feminine feeling and to layer the space.
In order to achieve the design goal, the design style, color and decorations are mostly related to 1930s Shanghai era. The designer had also set up the newest smart home system to be technology modern within the retro Shanghai theme space.

Eric Tam

从事室内设计行业已经超过 20 年。拥有一家属于自己的设计公司一直是 Eric Tam 的梦想，他果断放弃了很好的工作机会移居澳门创立了昊设计工程公司。经过多年的努力，Eric Tam 及他的昊设计硕果累累，不仅获得许多奖项还加入了不同的协会，备受各界人士的认可。

Eric has been involved in the Interior Design Industry for over two decades. Eric always had a desire to open his own firm, he eventually left the prior company to pursue and eventually establish ET Design & Build ltd. He and ET Design have received numerous awards and joined different kind of associations.

让音乐奏出空间
Let the music do the talking

项目地点：中国香港 / Location :Hong Kong, China
项目面积：896 平方米 / Area : 896 m²
公司名称 / Organization Name : Artwill Interior Design House

IAI 设计优胜奖

Artwill Interior Design House

Artwill Interior Design House于2002年成立，屡获殊荣，专门为家居提供美感与实用性并重的设计方案，实现客户理念。Artwill团队具有丰富经验，并由设计总监Regina Kwok领导。Regina拥有18年室内设计经验，曾在中国香港、澳洲、加拿大及美国完成多个项目，在全球均备受认同，曾荣获40 UNDER中国（香港）设计杰出青年、亚太区室内设计大奖（APIDA）及中国好设计等多个奖项。

Founded in 2002, Artwill Interior Design House is an awards-winning firm specializing in design solutions for your home that combine aesthetics and practicality to realize the visions of their clients. Heading a highly experienced team, creative director Regina Kwok brings 18 years of interior design experience and has delivered projects in Hong Kong, Australia, Canada and the US. Her projects have been recognized world-wide, winning multiple awards including China Interior Design 40 Under 40 (Hong Kong), China, 24th Asia Pacific International Design Award (APIDA) and China Good Design Gold Winner.

设计师的灵感源自客户对音乐的热爱。有趣和异想天开的元素使这个家庭像一个音乐录音室。走廊隐藏的门创造了一个白色的画布，用于展示主人的电影海报。移步起居室，主人的令人印象深刻的音乐控制台迎接您的到来，这是这个空间的核心。凸起的工作平台将主人置于生产者座位上，为他的收藏提供了充足的货架空间。设计师使用灯光营造温馨有趣的氛围。沿着天花板的环境照明投射出温暖的光芒。餐厅上方的美学灯带来一种异想天开的元素。落地窗可以自然采光。即使在卧室里，设计师也会使用折叠式床头板，在收回时可以让额外的光线透过。

The designer' s inspiration comes from the client' s love of music. Fun and whimsical elements make this home resemble a music recording studio.Concealed doors in the entrance hall creates a white canvas for displaying the owner' s movie posters. Moving into the living room, you are greeted by the owner' s impressive music console, the centerpiece of this space. A raised working platform puts the owner in the producer seat and provides plenty of shelfing space for his collection.The designer plays with lighting to give the space a warm and fun feel. Ambient lighting along the ceiling casts a warm glow. The aesthetic lamp above the dining room brings in a whimsical element. Floor to ceiling windows allows in natural light; even in the bedroom, the designer uses a folding headboard, allowing additional light to shine through when retracted.

张宅
Residence Zhang

项目地点：中国湖州 / Location : Huzhou, China
项目面积：70 平方米 /Area : 70 m^2
设 计 师：曾濬绅 / Designer : Jyun Shen Zen

IAI 设计优胜奖

曾濬绅 Jyun Shen Zen

曾濬绅对于室内设计领域、软装设计领域，皆有着透彻的了解及运用。深知设计不单只为室内设计，更应该是一种对生活观察之大数据结晶。并且将此大数据，运用在各种设计当中。

Jyun Shen Zeng has a thorough understanding and application of interior design and soft design. I know that design is not only for interior design, but also a kind of big data crystallization for life observation.
And use this big data in various designs.

此户仅为70平米，极小户型，但却利用了各种设计手法，使空间变得更大、更简洁。

1.玄关延伸到客房的电视墙，加强了空间的广度。2.通透的书房，使客厅空间更加深入。3.磁砖木地板的无缝拼接，扩大使用的想像。4.黑色出风口延伸各个空间，各空间串联，加强空间放大效果。5.灰色墙面配合白色天花板，使空间高度无形向上延伸。6.床头及天花板线条延伸，让空间像是加高加长。7.黑灰白为基底色，搭配些许金色提升高贵感。

This space is 70 square meters and belongs to a small space.But a lot of design ideas are used to expand the space.
First:The entrance extends to the TV wall of the room, enhancing the width of the space.Second:Transparent studio makes the living room deeper.Third:The seamless combination of tiles and wooden flooring expands the imagination of use.Fourth:A black air outlet extends each space and each space is connected in series to enhance spatial amplification.Fifth: Gray walls and white ceilings make the space higher.Sixth: The extension of the headboard and ceiling lines makes the space look like extension.Seventh: Black and gray and white are basic colors and a small amount of golden yellow to enhance the sense of nobility.

中场休息
Intermission

项目地点：中国湖州 / Location : Huzhou, China
项目面积：1200 平方米 /Area : 1200 m^2
公司名称：创空间集团 / Organization Name : Creative Group

IAI 设计优胜奖

創空間
CREATIVE GROUP

汇集建筑与室内设计背景的多元人才，配合组织化管理模式，将工程进度做系统化的全面控管，提供从前期沟通、设计、合约拟定到后期执行的完整服务。

2000年成立"权释设计"2014年，2015年陆续成立"CONCEPT北欧建筑"与"JA建筑旅人"建筑设计公司，传递人与自然和谐共生的理念精神，秉持着"提升国人生活品质、共创生活美学体验"为目标持续优化。

Bringing together diverse talents in the background of architecture and interior design, and coordinating the organizational management model, systematically and comprehensively control the progress of the project, providing complete services from pre-communication, design, contract formulation to post-execution. In 2000, the company established the "ALLNESS Design" in 2014. In 2015, it established the "CONCEPT 北歐建築 " and the "Journey Architecture" architectural design company to convey the spirit of harmony between Humanity and Nature Environment, and uphold the "quality of life of the people." is the continuously optimized goal .

"Intermission"，专指戏剧表演的中场休息，表演者抽离角色后的重整与放松，调整状态再次迎向下一幕的展开。呼应业主表演工作者的身份，以剧场黑幕、布景及灯光元素，将公区的场域打破，配合L型空间的轴线，让每个区域都有属于自己的一幕景，彼此连接又能合为舞台剧的布景概念；透过深色低彩度调性及反射性材质展现创造出中场休息般的沉静居所，寻求能与自己深度交谈的空间；让心灵沉淀，让自我展现。

INTERMISSION, specifically refers to the break time during the theatre performances, our client as a performing artist ,design elements came from theater's set and lighting. The L-shaped axis designed to break the rules and make different areas connected fluently . Just like a stage play scnee ; through the dark low-color tonality and reflective material to create a quiet place for client's soul be calm and take rest easily at home.

年轮说
Living Wild

项目地点：中国台湾 / Location : Taiwan, China
项目面积：27 平方米 / Area : 27 m²
公司名称：创空间集团 / Organization Name : Creative Group

IAI 设计优胜奖

进入玄关，映入眼帘的是斜屋顶设计与原始混凝土质感的结构梁，具有山中小屋的意象，搭配铺设笔直的磨石子地砖，一路延伸到窗边，呼应窗外震撼的天然美景，位处置高点，天气好的时候还可以眺望海洋。客餐厅主空间以大面积的不锈钢、木皮材质为主，并以充满绿意的植栽跟软件配置为辅搭配，呈现出“刚柔并济”的设计功力。设计师运用纯粹的材质，在比例与视觉上成就简练的空间。餐桌上的订制麻绳吊灯打破原有的协调秩序，呈现了原野的粗犷不羁，让空间增添几分丛林感。本案空间中另一亮点即是由玻璃墙所隔成的多功能房。以玻璃作为隔间墙面，不仅引入大量光线，同时放大空间的视觉效果。拉下铝百叶，亦可享受兼具隐私感的空间。

Entering the porch, we can see the sloping roof design and the original concrete texture of the structural beams, which have the image of a small house in the mountains, with the laying of straight grindstone floor tiles, extending all the way to the window, echoing the natural beauty of the shock outside the window, disposing of the high point, when the weather is good, we can also look at the sea! The main space of the guest restaurant is mainly made of stainless steel and wood, and is complemented by green planting and software configuration, showing the design power of "combining rigidity and softness". Designers use pure materials to achieve concise space in proportion and vision. The customized hemp rope chandeliers on the dining table break the original coordination order, showing the rough and unruly nature of the field, which adds a little jungle sense to the space. Another bright spot in the case space is the multi-functional room separated by the glass wall. Using glass as partition wall not only introduces a lot of light, but also enlarges the visual effect of space. Pull down the aluminum louvers and enjoy the privacy of the space.

序列
Transition

项目地点：中国湖州 / Location : Huzhou, China
项目面积：1200 平方米 /Area : 1200 m²
公司名称：创空间集团 /Organization Name : Creative Group

IAI 设计优胜奖

设计师将空间美感塑造为第二层质感的基底，让居住者及艺术摆设品作为空间，第一层视觉焦点。以层序为设计理念，在山岚之中配搭环境不同的层序变化，层层堆叠出夫妻俩的质感生活。公领域以简单纹理的浅色地砖作为背景色彩，衬托出电视墙使用的蒙马特灰仿古面大理石，并运用切割面作出如隔栅般的线条变化，搭接立面细致线板，勾勒出简约却精致的艺术氛围。

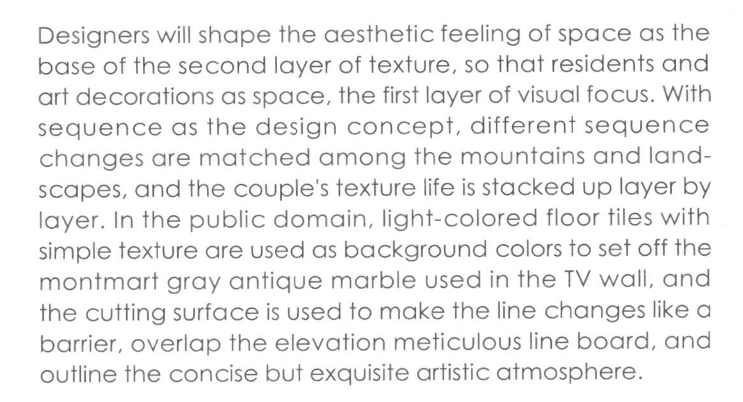

Designers will shape the aesthetic feeling of space as the base of the second layer of texture, so that residents and art decorations as space, the first layer of visual focus. With sequence as the design concept, different sequence changes are matched among the mountains and landscapes, and the couple's texture life is stacked up layer by layer. In the public domain, light-colored floor tiles with simple texture are used as background colors to set off the montmart gray antique marble used in the TV wall, and the cutting surface is used to make the line changes like a barrier, overlap the elevation meticulous line board, and outline the concise but exquisite artistic atmosphere.

方圆之间
Square and Round

项目地点：中国台湾 / Location : Taiwan, China
项目面积：27 平方米 / Area : 27 m^2
公司名称：创空间集团 / Organization Name : Creative Group

IAI 设计优胜奖

在方与圆、白与灰之间穿梭，以人为本，将空间适度留白，保有空间弹性，将装饰降至最低，让空间简洁有力，而隐约地利用软件去引导和延伸人的视线，进而品味室内外的景色，相互映照。将原有隔间打破，重组空间轴线，让视线开阔而延伸至窗外与景色产生连接，透过隐藏门片的开闭，让宠物与空间保有弹性，柜体虚实的线条，圆弧形的天花，企图在进入厅室后，创造出无垠的视觉层次。极简的纯净中，取用树木宽广盘踞又包容无垠的意象，生长出土地的气息。在家里打造舒适的角落，享受日常生活中片刻的美好，与宠物共享温馨时光。

Shuttle between square and circle, white and grey, people-oriented, moderate white space, space flexibility, decoration to the lowest, so that the space is simple and powerful, and the vague use of software to guide and extend people's vision, and then taste the indoor and outdoor scenery, mutual reflection. Break the original compartment, reorganize the space axis, make the line of sight open and extend to the outside of the window to connect with the scenery. Through the opening and closing of hidden doors, let pets and space keep elasticity, empty and solid lines of cabinets, circular ceiling, in an attempt to brew after entering the hall, create an infinite visual level on a single cold interface to derive, in the simplicity of purity, take broad use of trees. Cohesion and tolerance of boundless imagery, the growth of the atmosphere of the land. Create comfortable corners at home, enjoy the moments of everyday life, and share warm time with pets.

信仰蓝图
Faith

项目地点：中国湖州 / Location : Huzhou, China
项目面积：29 平方米 /Area : 29 m^2
公司名称：创空间集团 /Organization Name : Creative Group

IAI 设计优胜奖

《圣经》中写道：“信心没有行为是死的。”所以设计师将屋主心目中信仰的蓝图，完美地呈现在这个项目中。

本案的屋主希望透过空间呈现他的信仰蓝图跟概念，所以他将信仰中的语言设计在空间里面，同时将他信仰中乐于分享的概念，也放了进去。

在表达室内设计理念中，相较于室内的美观，他更重视信仰的传递，核心是：信心。同时也将屋主爱旅行的个性呈现在空间之中。

"Confidence, no action is dead" - the Bible. Therefore, the designer presents the blueprint of the belief in the owner's mind perfectly in this project.
The owner of the case wants to present his belief blueprints and concepts through space, so he designed the language of his beliefs in space and put the concept of his belief in sharing.
In expressing the interior design concept, he pays more attention to the transmission of faith than the aesthetics of the interior. The core is: confidence. At the same time, the personality that the owner loves to travel is presented in the space.

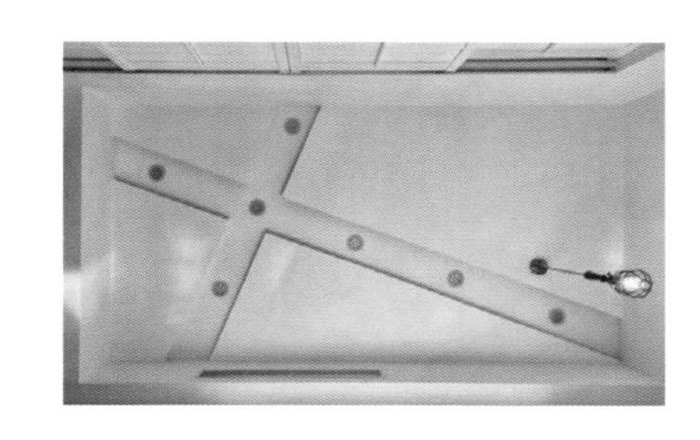

体现空间舞动与设计
The Interpretation of Space & Movement

项目地点：中国香港 / Location : Hong Kong, China
项目面积：1984 平方米 /Area : 1984 m²
公司名称 / Organization Name : Artwill Interior Design House

IAI 设计优胜奖

Artwill Interior Design House

Artwill Interior Design House 于 2002 年成立，屡获殊荣，专门为家居提供美感与实用性并重的设计方案，实现客户理念。Artwill 团队具有丰富经验，并由设计总监 Regina Kwok 领导。Regina 拥有 18 年室内设计经验，曾在中国香港、澳洲、加拿大及美国完成多个项目，在全球均备受认同，曾荣获 40 UNDER 中国（香港）设计杰出青年，亚太区室内设计大奖 (APIDA) 及中国好设计等多个奖项。

Founded in 2002, Artwill Interior Design House is an awards-winning firm specializing in design solutions for your home that combine aesthetics and practicality to realize the visions of their clients. Heading a highly experienced team, creative director Regina Kwok brings 18 years of interior design experience and has delivered projects in Hong Kong, Australia, Canada and the US. Her projects have been recognized world-wide, winning multiple awards including China Interior Design 40 Under 40 (Hong Kong), China, 24th Asia Pacific International Design Award (APIDA) and China Good Design Gold Winner.

随性自如的生活尤如舞蹈，与生活在静态的建筑里，两者的共通之处承载艺术，体现对空间的解读。

设计师活用形、线及高低落差，以直线、斜面及几何贯穿全屋，营造动态，让住户犹如舞者，亲历主体与空间的交织流动。

天花融入阶梯结构，既加强空间感，亦让视觉从饭厅渐渐伸延至客厅。

此家居活用不同设计元素，带出空间及动态的体现。

Dance is motion while architecture is static, but the two disciplines of art share a commonality – the interpretation of space.

The objective of dance is to use movement in a meaningful engagement between body and space. Similarly, the designer hopes the inhabitants can experience the visual, sensory, and motion of the space they live in.
To create motion, the designer plays with shapes, lines, and elevation - vertical lines, slanted planes, and tall rectangles are combined to accentuate movement and energy.

The challenge comes from the different ceiling heights. To enhance the spatial awareness, the designer incorporated steps in the ceiling that gradually leads you from the low dining room ceiling to the tall height of the living room.

华发水岸度假屋
Huafa Waterfront Holiday House

项目地点：中国广东 / Location : Guangdong, China
项目面积：142 平方米 /Area : 142 m²
设 计 师：李友友 / Designer : Evans Lee

李友友 Evans Lee

李友友毕业于南京大学及中山大学 MBA 工商管理硕士，进修于中国美院现代设计高研班及亚太酒店设计高研班，于 2015 年开展个人事业，成立李友友室内设计有限公司。

Founder Mr. Evans Lee graduated from NTU and Sun Yat-Sen University with an MBA in Business Administration. He studied at the China Academy of Art Modern Design High School and the Asia Pacific Hotel Design Institute. He started his career in 2015 and established Evans Lee Interiors Limited.

空间设计以白蓝色为主，除了中央空调、投影机和射灯，天花板一片干净的白色，没有多余的装饰。室内空间的墙身是由等间距的竖向浅灰色特殊材料组成，视觉上隐藏了转角位、走廊和通道门口，看起来更加开阔，增强空间的整体感。木地板颜色偏浅，成为调和空间的暖色调。

客厅的大幅挂画起到提升客厅艺术氛围的作用，挂画面向阳台，纯粹的蓝色在自然光底下显得更有质感和层次。客厅地毯也是特别订造，贯彻极具比例元素的设计，地毯为不同程度的蓝色，极具质感。

儿童多功能房中靠窗放置的沙发善用自然光之余，亦可于休息的同时欣赏窗外景致。在物料的运用上，运用鲜艳的颜色和不同质感的布料，令室内环境鲜明活泼。

The space design gives priority to with white blue, besides central air conditioning, projector and shoot a lamp, the ceiling a piece of clean white, do not have redundant adornment. The wall body of indoor space is made up of the special vertical and light gray materials with equal space. The corner position, corridor and doorway are hidden visually. Wooden floor color slants shallow, become the warm color of harmonious space.
The effect that the large hang picture of the sitting room raises sitting room artistic atmosphere, hang a picture to the balcony, pure blue appears to have simple sense and administrative levels more below natural light. Sitting room carpet also is special order make, carpet is different level blue, have simple sense extremely.
The sofa that muti_function room relies on a window to place is good with natural light besides, also can appreciate the scene outside the window while resting. In the use of materials, the use of bright colors and different texture of the fabric, so that the indoor painting is bright and lively.

十三步走完的家
A House within Thirteen Steps

项目地点：中国上海 / Location : Shanghai, China
项目面积：35 平方米 /Area : 35 m^2
公司名称：立木设计研究室 / Organization Name : L&M Design Lab

IAI 设计优胜奖

立木设计研究室
L&M design lab

立木设计 L&M Design Lab

L&M 隐喻 Logic & Magic，立木以理性为基石，积极拥抱复杂的设计条件与需求，并总能在这个基础上呈现充满魔力的作品。立木坚持大胆的设计构思和深入的学术研究，业务涵盖建筑、景观、室内、机电，立木以系统的工作流程和优质的设计服务为基础，立足于高速发展与迭代的时代，以超越的姿态为人们提供更优质的空间媒介。

L&M metaphors Logic & Magic, L&M is based on rationality, actively embracing complex design conditions and needs, and always presents magical works on this basis. L&M adheres to bold design ideas and in-depth academic research. The business covers architecture, landscape, interior, and electromechanical. Based on systematic workflow and high-quality design services, L&M is based on the era of rapid development and iteration. People provide better space media.

该项目是东方卫视《梦想改造家》节目作品。项目位于一幢老公房顶层，室内面积不足35平米，无法满足一家五口的日常需求。设计巧妙借用“翠玲珑”的空间原型，通过“入口-外厨-老人房”“内厨-外厨-客厅”两个“翠玲珑”巧妙拓展空间，可变家具和毛细血管的应用，也显著提升了老公房的室内品质。厨房与老人房的对景窗、客厅电视墙的张掖丹霞、世界上最小的歌唱厅等都充满人文关怀。这其中一户极小住宅的改造，在熙熙攘攘的人间烟火中营造了一个浪漫诗意的梦想家园。

The program of Oriental TV's dream reformer. The project is located on the top floor of a old public house with an indoor area of less than 35 square meters, which can not meet the daily needs of a family of five. The design ingeniously borrows the "Cui Ling Long" space prototype, through the "entrance - outside kitchen - elderly room" and "inside kitchen - outside kitchen - living room" two "Cui Ling Long" ingeniously expands the space, variable furniture and capillary application, but also significantly improved the indoor quality of the old public room. The opposite windows of the kitchen and the old people's room, the living room TV wall of Zhangye Danxia, the world's smallest singing hall are full of humanistic concern. The renovation of one of the smallest houses has created a romantic and poetic dream home amid the hustle and bustle of human fireworks.

四胞胎跑道之家

Four the Home of the Runwa

项目地点：中国上海 / Location : Shanghai, China
项目面积：60 平方米 /Area : 60 m²
公司名称：立木设计研究室 / Organization Name : L&M Design Lab

IAI 设计优胜奖

该项目是东方卫视《生活改造家》作品。改造前"东方明珠"四胞胎一家挤在60平方米的毛坯房中，空间杂乱拥挤。改造设计的灵感来自黄浦江，通过一条南北贯通的黄色跑道联系所有空间，更暗示了这家人和上海的奇妙缘分。设计师将不常用的功能折叠，并放大承载着家庭大部分活动的公共空间，随物赋形的墙充满童趣，双操作台的厨房和双卫生间提升效率，同时巧妙布置下了男孩、女孩和父母的独立房间。满足生活必须之余，设计师利用视线引导和尺度变化，营造出日常生活的趣味性。

The works of Oriental TV's life reformer. Before the transformation, the "Oriental Pearl" quadruplets was crowded in 60 square meters of rough housing, and the space was crowded. Inspired by the Huangpu River, the renovation connects all the spaces through a north-south yellow runway, suggesting the family's wonderful relationship with Shanghai. The designer folds out the unused functions and enlarges the public space that carries most of the family's activities. The walls are filled with children's interest. The kitchen and toilet with two operating tables improve efficiency. Meanwhile, the independent rooms for boys, girls and parents are arranged ingeniously. To satisfy the needs of life, designers use visual guidance and scale changes to create a daily life of interest.

往日情怀
The Way We Were

项目地点：中国台湾 / Location : Taiwan, China
项目面积：120 平方米 /Area : 120 m²
设 计 师：吕沛杰 /Designer : Lü Peijie

IAI 设计优胜奖

吕沛杰 Lü Peijie

居间国际室内装修设计执行总监，设计师的立场非常单纯就是提供“观念和方法”，“观念”就是提倡一种生活态度与生活方式，“方法”是以设计为本位。“核心价值”就是做对的事。基于以人为本的设计哲学与初衷来因应每个案件。创造让空间、自然与生活行为环环相扣，让人们适得其所。

PEI CHIEH LU, Execution Director of the InSpace Interior Design Ltd. The standpoint of PEI-CHIEH is very simple, one of which provides "Concept and Strategy" the "Concept" advocates improving the living attitude and lifestyle. The "Strategy" is basing designs on original ideas. The core value is doing things right for the result to fulfill its value. Base on human-oriented design philosophy as the original intention to respond to each project. They design the space, where the space, nature, and living behavior are closely linked and appropriate.

人们生活在钢筋水泥的丛林里，在忙碌的每一天后都需要释放一些压力，这个空间重现了客户的生活方式，设计师特别重视细部设计，采用大量的木质纹理代替奢华的光泽材料，营造出温馨古朴的英式酒吧风格。串接在一起的公共空间在早上是儿童游乐场，在夜晚和周末则成为家庭中心和朋友间的交流中心。

People need to release stress after a busy day especially when living in the reinforced concrete jungle city. This space represents the lifestyle of the client, designer focus on the detail design of using massive wooden texture instead of luxury glossy materials to create a warm quaint British pub style. The connected public area is kids' playground in the morning, it becomes families and social intercourse center at night and the week-end.

源点生活
A Simple Life

项目地点：中国台湾 / Location : Taiwan, China
项目面积：196 平方米 / Area : 196 m²
公司名称：树屋室内装修设计有限公司 / Organization Name : Tree House Design

IAI 设计优胜奖

树屋室内设计
Tree House Design

树屋设计是一支由梦想与执行力组成的设计团队，抱持着对美学的绝对坚持、独特的创意发想，针对居者的需求与个人生活习性，展现细腻的关怀与深度的美学是创办人坚持的理念，辅以精确扎实的施工品质，为业主打造一件件隽永而能温暖人心的作品。

Tree House Design Is A Team Of Young Designers, Drawing Creative Thoughts From Our Aesthetic Point Of View With Fine Execution. We Believe Our Design Takes Consideration Of The Needs Of Our Residents And Their Styles Of Living, Which Reflects Through The Details Of Precise Construction. By Combining What Our Clients Want And Our Aesthetic Concept, Our Timeless State-of-the-art Design Will Always Be The Home To Return To.

客厅与书房采用开放式格局，延伸出宽敞的舒适歇憩空间。餐厅的线性吊灯与大面积手工拼凑的几何墙面造型巧妙呼应，搭配造型简约暖色原木长桌，相异的序列感让用餐环境更具张力。卧室及浴室则以沉稳灰色基调，勾勒出精致宜人的寝寐空间。起居室引入自然光佐以温润的木质点缀，呼应整体悠然惬意的居家韵味。以净简的设计手法拒绝浮华繁复，营塑质感温暖的居家氛围。家，是远离世俗喧嚣的港湾。我们想要回归更精简舒适的生活型态，我们想要去芜存菁，最后留下的线条刻画出生活精致美学与深度。

The wood color tone was added into the black, white, gray color tone, it strengthens the temperature and layers without loosing texture. There are a lot of lines in the details which stretches the sense of space and distinguishes the items in space, and also allows them to pop out from their surroundings.
The whole arrangement is styled with a sense of openness. The living room is connected with the study room, which, opens up the horizon and also widens the areas available for comfort and leisure. The dining room is, decorated with hanging lights, and the floor made of granite rocks with geometric patterns. All these things in the dining room correspond with the theme, and the lone wooden table,also conserves the warmness of the space. The style of the bathroom is enhanced by a gray color tone, which allows it to become more sophisticated, and creates a calm atmosphere. So in a simple way, we have created a warm space with texture.

浮尊宅
A Distinguished Style with Glory and Calm

项目地点 : 中国台湾 / Location : Taiwan, China
项目面积 : 100 平方米 /Area :100 m²
公司名称 : 上云空间设计 / Organization Name : Creative Design

IAI 设计优胜奖

上云设计 Creative Design

走在云端，做出超越想像的事。以细心、用心、贴心三心打造业主的信任。
上云，顾名思义为云的古字，希望“家”的感觉，如天上的云朵一般，无拘无束且悠游自在的感觉。很重视“人”的接触和感受。在与人互动之中，看空间的角度宛若一门精湛工艺，从纸上谈兵幻化为实境，打造每位屋主专属的居家空间，创造每位家中成员所想要的幸福。

"Going in the clouds and making things beyond imagination" builds the trust of the owners with "carefulness" "attentively" and "intimateness". Shangyun, as the name suggests, is the ancient word of the cloud. It hope that the feeling of "home", like the clouds in the sky, is unrestrained and comfortable. It attach great importance to the contact and feelings of "people." In the interaction with people, the angle of looking at the space is like a superb craftsmanship. From the paper, the soldiers are turned into reality and the exclusive home space of each homeowner is created.Create the happiness that every member of the family wants.

挑高6米的宽阔场域，突显出高耸的气势与格局。公共空间以电视墙为轴心，并以大理石、镜面材质衬托出浮华、尊贵的感觉；而前后两面由天至地直立木皮造型墙，带出内敛沉稳的氛围。互相交叠出一种反差，一种格格不入的互补效应，就像家人间的不同性格，交织出幸福的生活。大器、优雅、尊贵，令人沉醉其中。

A spacious area with a height of 20 feet - using TV wall as pivot of living room in order to express imposing layout and impression. Meanwhile, presenting the feeling of glory and dignity by materials of marble and mirror. Two top-to-end wooden crafted walls at front and back presenting a modest vibe.
Compensated with contrary, just as different personality among family members which build up a life with happiness.
Immersed in the vast, glorious and elegant vibe.

住宅办公合二为一的现代休闲好生活

Residence x Office A modern yet leisurely lifestyle

项目地点：中国台湾 / Location : Taiwan, China
项目面积：100 平方米 /Area : 100 m²
公司名称：上云空间设计 / Organization Name : Creative Design

IAI 设计优胜奖

“温润”“流动”“延伸”是整体视觉风格，串联出“放松”“愉悦”“隽永”的主题氛围。而动线以“环状动线”为主轴，随处都可坐可用，并无单纯的走道浪费，透明隔间和零走道十分通透。

木质占整体六成，水泥粉光三成；这两种朴实材质一温一冷，再加上大理石和少量镀钛，呈现出人文休闲的美感。以灰、白、木纹3种色彩为主轴，调配些许黑色点缀，使空间更为精致大器。

"Warmness", "Flow" and "Extension" are the overall visual styles. The theme of "relaxation", "pleasure" and "perpetuation" is connected in series. The moving line takes the "ring moving line" as the main axis, and can be used anywhere. There is no simple waste of the walkway, and the transparent compartment + zero walkway is very transparent.

Wood accounts for 60% of the whole, and cement powder is 30%; these two simple materials are warm and cold, plus marble and a small amount of titanium plating, showing the beauty of humanity and leisure. The three colors of gray, white and wood grain are used as the main axis, and some black embellishments are arranged to make the space more exquisite.

璀灿脉脉
Van Der Vein

项目地点：中国台湾 / Location : Taiwan, China
项目面积：131 平方米 /Area : 100 m²
设 计 师：翁新婷 / Designer : Weng Xinting

IAI 设计优胜奖

翁新婷 Weng Xinting

视觉传达艺术与建筑的学历背景，养成对于色彩与材质细节的极高敏锐度，善于解构流行趋势与多样元素，借材料与软装的运用，发展细腻的人际与空间互动关系，诠释机能与美学并具的空间。

Visually convey the academic background of art and architecture, develop a high degree of sensitivity to color and material details, be good at deconstructing fashion trends and diverse elements, and develop delicate interpersonal and spatial interactions through the use of materials and soft clothing, interpreting functions and The space of aesthetics.

如此光彩夺目的屏息，是石与金工激荡的共存，是空间内化后的璀璨主义。展开的大理石并非富丽拼花与对纹雕饰，浑厚粗旷的石纹脉络在线性的排列下展现岁月的斑斓，云白银狐的延展清新淡雅、细腻生动。转折过后，借着几何形的形体组合，突破单向垂直与水平的发展。精细的金属以弦线化作一扇屏风，微乎其微却注目的光泽，隐晦之间融入石材脉络原始的瑰丽。不锈钢原色于内退空间，虚实之间表态不失简敛。午后斜阳穿透金色薄纱洒入室内，空气中弥漫缕缕闪光，随影摇曳，光灿中璀彩奔放，脉脉不得语。

The abode is an eclectic extravaganza of natural marble paneling. Without too sumptuous measures, the space accentuates the innate elegant essence of rock veins comparing to metallic modernity. GrigioCarnico is gracefully arranged in linear paneling forms, while white Cherokee marble transforms its vein irregularity into chevron pattern, pure yet lively. Titanium color-gilded stainless trims delineate lavish modernity along the way. While sunlight shines through the curtain sheer, the space generates glamorous shimmers, culminating with luxury yet serene vibes for residence.

霏雾烟波

Fog Floated

项目地点：中国台湾 / Location : Taiwan, China
项目面积：113 平方米 /Area :113 m²
设 计 师：翁新婷 / Designer : Weng Xinting

IAI 设计优胜奖

无尽的墨晕纹理深沉却轻薄，浮云流脉般的渲染，淡然且悠暗，金属壁饰的镜面属性更为显著，亦与大理石的天然瑰逸互相契合，空蒙如轻雾，东方意境弥漫于极简的空间维度。随视觉的推进，材质层次与空间次元进而平衍共生；灯源光晕蒙蒙回荡，不锈钢与漂流木纹的高冷基调乍现暖韵；透光灰玻轻盈的通透性如湖为镜，日光弄影随百叶上下摇曳。一日之间，空间与居住随光的转换若影若霰、忽隐忽现，神秘的深邃令人着迷。

Entering the abode, polished stainless wall decor attracts attention, not only because of its glow, but also of its contrast with backdrops of ink-wash cloud textured in deep color hue, which is unintentionally coherent to natural veins from marble wall, bringing misty and airy sensation, composed and profound. Orient vibes linger in a contemporary space. Drift to open-concept area, dim lights blur through grey transparent glass, propelling to progressive interaction with spaces and materials. Lens flares illuminate metallic glows, and give a warm tone to salvaged-wood grains. Sunlight sways through shutter blinds, flickering in hazy space, culminating with calm yet fascinating ambiance.

静若觅谧
Virtual Reality

项目地点：中国台湾 / Location : Taiwan, China
项目面积：211 平方米 /Area : 211 m²
设 计 师：翁新婷 / Designer : Weng Xinting

IAI 设计优胜奖

踏出梯厅，映入眼帘的是大理石与皮革交互点缀的轻奢，悬浮柜体腾空，使梯厅有别于以往的入口，前后层次亮暗的转折，一侧的洞石质感温厚，引着余光前进，透彻如雪的银狐端景，延续另一侧镜面交互作用，似空间延伸幻化，光影透露线索，隐约将入口藏匿。圆弧向下延伸聚焦，简单流畅、外方内圆，日光熠熠，清透的玻璃以不同面貌拼凑，波光粼粼。灰褐色系使氛围沉稳内敛，空间随光影而转折不同特性。

Entering the hall, coexistence between marble and leather gives sumptuous vibes. The hanged cabinet gives a newness to the space. Multiple layers between back and forth, and the travertine wall is warm. The other side of marble wall continues the mirror interaction on the other side, like the spatial extension of the illusion. The light and shadow reveal clues, vaguely hiding the entrance. The arc extends downwards to focus, simple and smooth, the outer circle is round, the daylight is shining, and the clear glass is pieced together in different faces, sparkling. The taupe makes the atmosphere calm and restrained, and the space changes with the light and shadow.

微古典现代白
Light Classical Modern White

项目地点：中国台湾 / Location : Taiwan, China
项目面积：149 平方米 /Area : 149 m²
公司名称：宸舍室内装修设计有限公司 / Organization Name : Chenshe Interior Design

IAI 设计优胜奖

宸舍设计
Chenshe Interior Design

空间因为人而具有生命力，而设计理应秉持着以人为本的理念，以舒适与机能为空间的核心。宸舍设计创立于 2007 年，从事之设计领域涵盖住家空间、商业空间、办公空间、大型聚会场所与建筑设计，创立十年以实作经验实践想像中的空间，以使用人为设计出发点。在人与美感和机能当中取得平衡，让使用者在空间里能够感受到空间美感所带来身心灵的愉快以及感动。

Space has vitality because of people, and design should adhere to the people-oriented concept and the core of comfort and function.
Chan She Design was established in 2007, Chance Design fields cover residential space, commercial space, office space and large gatherings. Site and architectural design, created a decade of practical experience to practice the space in the imagination, to use the artificial design starting point
Balance between people and beauty and function
Let the user feel the pleasure and touch of the body and mind brought by the space beauty in the space.

空间打造，最重要的是贴切居住者的所需。而宅装，更加深空间视感与质地。透过设计师的巧思融汇，将业主最喜爱的轻古典风韵，展演在大面积的净白基调中。并融入丰富有序的收纳规划，呈现出微古典并合着现代白的层次逻辑。在引借宽阔大景的高楼层宅邸风华里，映透出内外相引借景的无价广阔。

以现代格局线条在框线与天花间划出界定，佐以微古典轻妆铺陈其中。甫入玄关，映入眼帘的是雕琢精巧的威尼斯镜，右面长落地推拉门柜可于进出间容大量收纳，呈现小型衣帽间型式。

The first priority of interior design is giving what the client needs indeed. For the client favoring most in light classical aura, the designer specifically intensifies the space characteristics of both sense of sight and texture in the capacious and white house. Meanwhile, the comprehensive storage spaces strengthen the merged quality of light classical and modern white that constitute the accordant relationship between the indoor and outdoor space.
The modern lines define the layout associated mildly with classical characters. The porch is decorated with the delicate Venice mirror, and the storage space behind the sliding door on the right is a small cloakroom.

积累之间
Inter-Accumulation

项目地点：中国台湾 / Location : Taiwan, China
项目面积：60 平方米 / Area : 60 m²
公司名称 / Organization Name：坊华室内装修设计有限公司

IAI 设计优胜奖

FH·Design

坊华设计　FH Design

坊华室内装修设计有限公司成立于 1998 年，拥有 20 年以上的工程经验，对施工品质能精准掌握，在设计层面上更注重不同空间的舒适、创意及实用价值。近年来，我们积极参与各大国际设计盛事，多组设计作品荣获金、银等大奖的肯定。我们擅用天然材质装饰空间，融入艺术与文物，感受自然的呼吸与律动，享受美学与生活间的共鸣。

Founded in 1998, Fanghua Interior Decoration Design Co., Ltd. has more than 20 years of engineering experience, and can accurately grasp the construction quality. At the design level, it pays more attention to the comfort, creativity and practical value of different spaces. In recent years, we have actively participated in major international design events, and many sets of design works have won the recognition of gold and silver awards. We use natural materials to decorate the space, blend art and artifacts, feel the natural breathing and rhythm, and enjoy the resonance between aesthetics and life.

在新旧蓝图之间，在手感与岁月间，以装置美学，赋予空间生命。在动线行径中，领略材质的巧夺变化，如装置艺术铺陈于其中，散发着岁月的洗练气韵，使空间价值趋于无形之上，于空间故事中找到经典的意涵。积累的学问，不仅只于新质材的应用。空间看似舞台，质材化为装置，在装置中有新与旧的混搭对立，彼此间找到平衡与协调，呈现出深具对比的价值感。仿旧质地呈现着时光积累的岁月感，如同沙发前的旧木电视柜体，表面肌理满载岁月斑驳。在同侧的厨区与洗手间门面，则以大片仿旧的锈蚀钢皮，呈现出自然感的岁月表皮质地。

The decorative elements commingling the vintage and fashioned styles, which have been synthesized into the installation aesthetics, bring forth the life in the residential space. The inventive skills presenting the dexterity of material textures manifesting the artistry of installation arts reveal the elegant charms transitioning the space characteristics to a classy and higher level. The adroitness of commingling distinct compositions focuses not only the physical qualities, but also the intangible nature of space that amicably adjusts the old and modern features to the ultimate perfection. The designer adeptly decorates the house with the attractions of vintage styles, of which the antique surfaces of woods before the couch and the vintage-style sheet steels alongside the kitchen and bathroom elegantly present the ambience of nature.

国泰双玺
Grand View Castle

项目地点：中国广东 / Location : Guangdong, China
项目面积：270 平方米 /Area : 270 m²
公司名称：层层室内装修设计有限公司 / Organization Name : CCID

IAI 设计优胜奖

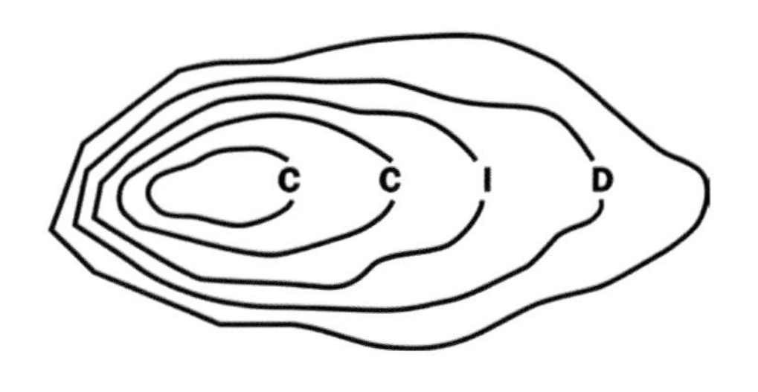

层层设计 CCID

室内设计是不同元素的搭配组合，透过与屋主的沟通，设计师将灵感、需求、美学、机能等置入空间。
重要核心在于设计的沟通，完工后样貌是屋主的期待，因此 3D 立体图的绘制，就是双方沟通中重要工具，以专业为有限的空间，创造无限的生命力。

Interior design is a combination of different elements. By communicating with the owners, designer manage to combine inspiration, demand, aesthetes, function and other elements into the space.
The important core is the communication of design. The appearance after completion is the expectation of the owner of the house. Therefore, the drawing of 3D stereograph is an important tool in the communication between the two sides.

精致景观豪邸，在闹中取静的佳美基地中孕育而生。眼观窗外山岚绿意畅快无限，近享自在空间质韵精巧细腻，注入多样层次细致的材质元素，以工艺等级的设计度量与功法，细腻雕琢于室。将原毛胚屋样式作整体脱胎换骨的呈现，将中西合并的东方风情与现代简约同时并置，打造出具艺术品般的传世家宅的格局。

设计底蕴如同层叠的积累，室内装置延续此层叠积累的概念，化为硬件间的精神意涵。如客厅沙发背墙上的抽象画作，描述山水间的积累关系。如餐厅天花的装置，在片层木质间的横向积累，透出均匀的灯源。

Surrounded by the great natural landscapes, the high-grade house decorations incorporating the exquisite materials and extraordinary craftsmanship show to the world an elegant quality of residential life. Overhauling the original layout, the designer concerning both the oriental delicacy and modern simplicity built a state-of-the-art legacy of residence.
The decorative adornments manifest the layered concepts, of which the art painting in the living room artistically simulates the natural landscapes, and the wooden-layered ceiling in the dining room brings out the cordial illuminations as well.

通北街

House Tongbei Street

项目地点：中国广东 / Location : Guangdong, China
项目面积：180 平方米 /Area : 180 m^2
公司名称：层层室内设计有限公司 / Organization Name : CCID

IAI 设计优胜奖

典雅温润，气韵柔美，满载女性优雅氛围，彰显空间气质，如穿梭于甜美梦境间。公领域以粉红跳色为主调，私领域以蒂芙尼经典蓝绿色用于主卧间，大面积白色基调中有着错落的配置与多层次惊喜色感，使空间线条在活泼与圆弧中散漫出典雅风华。

空间需求为体贴居者生活，配置室内机能，风格则以彰显视觉美学而演绎。空间领域讲究以层次顺序满足视觉逻辑，在顺序间引导出层次脉络，并以混搭风格丰富作品视觉。

以通透白为本质，配搭白色异材表皮作质感混搭。客厅电视主墙以进口卡拉拉大理石砖大面铺陈，砖面呈错置跳序，直向填缝处压金色线条亦呈现金色跳序，清新中隐约透出低调奢感的活泼。

The feminine ambiences, which is classic and graceful, bring out the joyful space characteristics in the house. The public area is primarily painted in pink; the master bedroom as the private area is mainly embellished with Tiffany Blue. The extensive white space, ornamented with the vivid colors, exposes the invigorating and elegant charms originated the lines.
The residential space furnished with reliable functionalities provides the pleasant life quality, while displaying the aesthetic lifestyle of in the meantime. The indoor space emphasizing the visual logic gives forth the multi-level sense, and the mix-and-match decorative style energizes the visual intensity as well.
The textures of materials consistently embellished with white exhibit harmonious charms. In the living room, the carrara TV wall surfaces adorned with golden lines diffuses the cheerful and delicate extravagant aura.

云上青
Cloud Castle

项目地点：中国广东 / Location : Guangdong, China
项目面积：200 平方米 /Area : 200 m²
公司名称：层层室内设计有限公司 / Organization Name : CCID

IAI 设计优胜奖

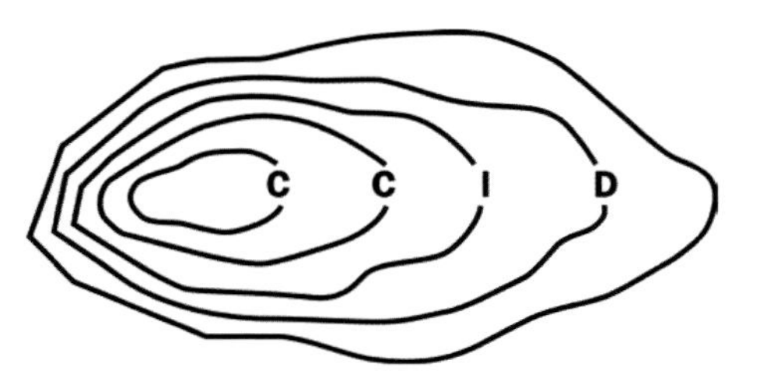

层层设计 CCID

室内设计是不同元素的搭配组合，透过与屋主的沟通，设计师将灵感、需求、美学、机能等置入空间。重要核心在于设计的沟通，完工后样貌是屋主的期待，因此 3D 立体图的绘制，就是双方沟通中重要工具，以专业为有限的空间，创造无限的生命力。

Interior design is a combination of different elements. By communicating with the owners, designer manage to combine inspiration, demand, aesthetes, function and other elements into the space.
The important core is the communication of design. The appearance after completion is the expectation of the owner of the house. Therefore, the drawing of 3D stereograph is an important tool in the communication between the two sides.

沉稳的木暮色，将空间感调和的深具层次。

以深木质调装点，辅以白色基调对比出空间感，在深浅明亮配比间，巧妙呈现空间动线的双重景致。并以色感区别出多重端景，设计师运用多样质材协调混搭，呈现空间美学的独特性。

回旋于玄关，从木隔栅间的穿透视感看出，隔栅边墙的大理石纹理呈横向千层，与地面沉色的手刮原木于厅大面积横向铺陈，再对应玄关边柜顶的大横梁，横越过双厅间一路至窗外大景前，此三方纹理线条感的横向铺陈，更为带动视感的延伸连接，增添空间中的气派延伸。

Steady dusk of wood harmonizes the sense of space at a deep level.
With deep wood decoration, complemented by white tone contrast to the sense of space, in the deep and bright ratio, cleverly presents the dual scenery of space dynamic line. And color sense distinguishes multiple landscape. Designers use a variety of materials to coordinate and mix, presenting the unique characteristics of space aesthetics.
Rotating around the entrance, we can see from the penetrating sense between the wooden partitions that the marble texture of the partition side wall is thousands of layers across the entrance hall, and the heavy hand-scraped logs on the ground are laid out horizontally in the hall, then corresponding to the large cross beam on the side cabinet top of the entrance hall, crossing the two halls all the way to the big view outside the window. The horizontal laying of the three texture lines is more dynamic, extending the connection of the sense of sight, and increasing the atmosphere in the space. Extend the pie.

一束
One Beam

项目地点：中国四川 / Location : Sichuan, China
项目面积：124 平方米 /Area : 124 m²
设 计 师：游世军 / Designer : You Shijun

IAI 设计优胜奖

游世军 You Shijun

2018 荣获 IAI 设计奖 住宅公寓优胜奖
2019 荣获国际空间设计大奖 Idea-Tops 艾特奖 公寓设计入围奖
2019 荣获 DNA 巴黎设计奖
2019 荣获第九届中国国际空间设计大赛（中国建筑装饰设计奖）银奖
APDF 亚太设计师联盟专业会员
IFI 国际室内设计师、室内建筑师联盟会员

2018 Awarded IAI Design Award Residential Apartment Excellence Winner
2019 won the International Space Design Award Idea-Tops Aite Award Apartment Design Finalist Award
2019 Received the DNA Paris Design Awards DNA Design Award
2019 won the Silver Award in the 9th China International Space Design Competition (China Architectural Decoration Design Award)
APDF Asia Pacific Designers Federation Professional Member
IFI International Interior Designer / Interior Architects Alliance Member

整体上，巧妙地利用镜子画作装饰，将各功能区连为一体，甚至是色彩和光影的搭配，也都融入了同样的自然色调，在现代气息与自然氛围中，穿梭自如。律动式切割石皮，在餐厅处打造一个通透、相融的功能区，定制的展示架上整齐地摆放着物品，唾手可得的便捷让等待的时光也变得美好。洗浴室的整体，选择了与整体和谐的黑白元素，并通过暗色系风格，纹理与材质的不同，让简单的色彩也拥有饱满的视觉感受。在洗浴室的里侧，设计师根据屋主需求定制了衣帽间，简约实用的储物架解决收纳需求，利用玻璃巧妙分割空间，在良好的采光与视野下，避免空间的杂乱，同时也让光线都可随之流动。

On the whole, the ingenious use of mirror painting decoration, the functional areas will be linked together, even color and light and shadow collocation, are also integrated into the same natural tone, in the modern atmosphere and natural atmosphere, shuttle freely. Rhythmically cut stone skin, create a transparent, integrated functional area in the restaurant, customized display shelves neatly placed items, handy convenience makes the waiting time become better. The bathroom as a whole, selected with the overall harmony of black and white elements, and through the dark color style, texture and material differences, so that simple colors also have a full visual experience. Inside the bathroom, the designer customizes the cloakroom according to the needs of the owner. Simple and practical storage rack solves the storage needs, cleverly divides the space by glass, avoids the clutter of the space in good daylighting and vision, and allows the light to flow with it.

幸运草
CLOVER

项目地点 : 中国福建 / Location : Fujian, China
项目面积 : 210 平方米 /Area : 210 m²
设 计 师 : 游小华、陈东升 / Designer : You Xiaohua, Chen Dongsheng

IAI 设计优胜奖

游小华 & 陈东升
You Xiaohua , Chen Dong-sheng

设计总监游小华和陈东升坚持认为 "在设计中做好一生"。设计不在于风格。首先，设计是为生活而设计的。更重要的是 将生活与空间美学完美结合。到目前为止，该公司已经赢得了许多国内室内设计奖项:Idea-Tops、Nest 奖、金堂奖、亚太设计奖。

Design directors You Xiaohua and Chen Dongsheng insisted that "doing a good life in design" . Design is not about style. First of all, design is designed for life. More importantly, the perfect combination of life and space aesthetics.
So far, the company has won many domestic interior design awards: Idea-Tops, Nest Award, Golden Hall Award, Asia Pacific Design Award.

凡是心中所喜欢的，看来总是最美。不囿于风格，在空间注入诗意的想象，让生活美学在设计中很好的体现。天汇设计总监说 "用一辈子的时间，做好设计" ，设计不在于风格，首先设计是为生活设计，其次是把生活和空间美学完美结合。好的设计不光是停留在视觉表象的美观，而是要通过设计发挥其可持续性、可调整性与互动性，并且拥有最大的灵活度，达到品质生活，达到人与空间的舒适相融。

Everything that I like in my heart seems to be always the most beautiful. Contrary to style, injecting poetic imagination into space, let life aesthetics be well reflected in the design. Tianhui Design Director said that "using a lifetime of time, designing well" is not about style. First, design is for life, and second is to combine life and space aesthetics. A good design is not only to stay in the beauty of the visual appearance, but to achieve its sustainability, adjustability and interactivity through design, and to have the greatest flexibility, to achieve quality life, to achieve the comfort of people and space.

你好，贝琪
Hi, Becky

项目地点：中国成都 / Location：Chengdu, China
项目面积：145 平方米 / Area：145 m²
设 计 师：钟莉 / Designer： Zhong Li

钟莉 Zhong Li

国内知名室内设计师，毕业于西南交通大学 ONE SPACE DESIGN 创始人。东京艺术大学进修生、创基金金 B 计划荣誉学员、米兰理理工大学艺术设计双硕士学位、香港室内设计师协会荣誉会员、40 UNDER 40 中国（四川）设计杰出青年、成都装饰协会会员、中国装饰协会同盟会会员、CCTV-2 交换空间常驻设计师、多乐士官方签约设计师、东方卫视就酱变新家特邀设计师。

A well-known domestic interior designer, graduated from the southwest Jiaotong University,founding of ONE SPACE DESIGN . Honorary student of Tokyo University of the Arts, Honorary student of Foundation B, Master of Arts and Design of Milan Polytechnic University, Honorary Member of Hong Kong Interior Designers Association, 40 UNDER 40 China (Sichuan) Design Distinguished youth, Member of Chengdu Decoration Association, Member of the China Decoration Association League of Decorative Associations, CCTV-2 Exchange Space Resident Designer, Dulux Official Signing Designer, and Oriental Satellite TV are specially invited designers.

"你好，贝琪"，这是印有她名字的、属于她和家人的刚刚好的样子。

单亲妈妈贝琪的家位于成都市区内一个普通小区。但只要一走进去，就会发现里面别有洞天：柜子是黑色的，墙纸是深灰的，甚至连纱帘都是炭黑色的。她还听从了风水大师的建议，在家里装了两个金色的马桶，每个来家里的朋友，都忍不住去坐一坐，感受一下……

这个设计，是设计师钟莉收集了大量贝琪的生活需求与她无数的交谈，把自己当成她来完成的一个设计，其目的就是想要让她拥有一个属于自己的空间，可以在这个空间里面放下她的所有强硬，可以放松地面对自己内心的柔软。

这些年钟莉一直在思考一些问题，设计师设计一个家的出发点是什么？其实就是创造一个最适合自己的居所，这里有自己的影子，有自己的气息，有自己的故事，不必在意别人的看法，因为日子都是自己过。在贝琪也愿意一同挑战的时候，就是设计师和客户双方都实现自我价值的旅程。

"Hi Becky", which the owner's name is imprinted into the project, belonged to only herself and her family.

The apartment is located in an ordinary neighborhood in downtown Chengdu. But as soon as you step into the apartment, the massive amount of black creates an unusual and mysterious atmosphere where you may lose yourself in it. Also, the owner has taken the advice of Feng Shui master and installed two golden toilets in each bathroom, thus making the guests more excited in exploring the space.

What is the starting point for designing for a homeowner? In fact, it is to create the most suitable place for the owner where they can identify and associate themselves with the space created; a journey and a challenge for both designer and homeowners.

Culture And Education Space

室内 —— 文教空间

我的秘密花园
My Secret Garde

项目地点：中国台湾 / Location : Taiwan, China
项目面积：40 平方米 /Area :40 m²
设 计 师：洪逸安 / Designer : Hong Yian

IAI最佳人文关怀大奖 IAI最佳设计大奖

业主愿意分享他们家的屋顶空间给朋友的孩子们，这个空间不但是儿童聚集的空间，同时也是可以做小型团体活动的地点。以一个有趣的多功能空间分享给社区是一件相当有意义而慷慨的事。

这是一个特别的空间设计体验，业主给了我们一个有滑梯的阁楼空间的想法，我们将这个有趣味的空间简单大方地实现出来。

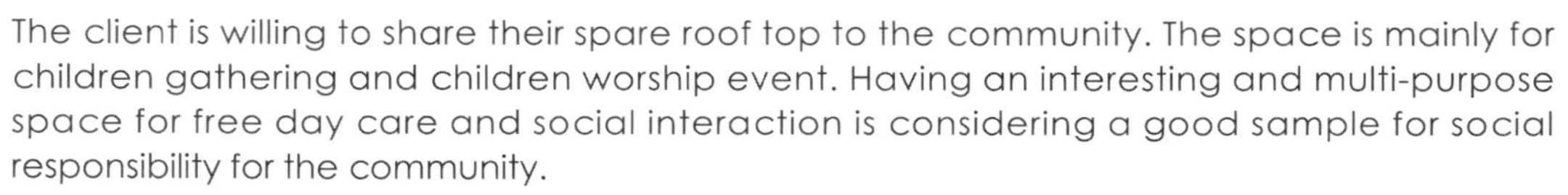

The client is willing to share their spare roof top to the community. The space is mainly for children gathering and children worship event. Having an interesting and multi-purpose space for free day care and social interaction is considering a good sample for social responsibility for the community.
It is a special spatial design experience, out client gives us an idea of a slide in a penthouse, and we recreate this interesting space in a simple and interesting manner.

洪逸安 Hong Yian

易设计主持设计师
中原大学设计学院博士生
中原大学室内设计设计硕士
中原大学室内设计学士

Yi design host designer
Phd student of Chung Yuan University School of Design
Master of Interior Design, Chung Yuan University
Bachelor of Interior Design, Chung Yuan University

CLC 北京
CLC Beijing

项目地点:中国北京 / Location : Beijing, China
项目面积 : 432 平方米 / Area : 432 m^2
设计师 : 大沽 · 日比野拓 / Designer : Taku Hibino

IAI最佳环境友好大奖

这是一个儿童保育支持中心，仅适用于位于北京市中心的发展中地区的会员，那里有许多高层公寓。会员的目标是住在这个地区的年轻家庭。它的功能不仅是儿童，父母和邻居也可以加入这个社区。在这个项目中，为了适应需要自由调整几个教育课程，每个教育课程都有各个模块。这就是我们设计的方式，以满足每个生活区域人们不同的需求，并适应一些灵活建设的现成条件。此外，很容易识别模块的每个用途。在这个设施中，我们把它作为关键概念："街头游戏在城市。"

This is a childcare support center only for members in the developing area located at the center of Beijing, where there are many high-rise apartments. Targets of membership are younger families who live in this area. It has the function that not only children but parents and neighbors can join this community here. In this project, as to adjust several education programs freely to suit the needs, each education program has each module. This is how we design to suit people' s needs which is different in each living area of users and suit some given conditions of building flexibly. Also, it' s easy to recognize each purpose of the modules. In this facility, we put it the key concept, "Street Play in City" .

大沽 · 日比野拓
Taku Hibino

日比野公司及旗下品牌"幼儿之城"幼儿设施项目的统筹负责人，KIDS DESIGN LABO 董事长。1972 年出生于日本神奈川县，毕业于工学院大学。他截至目前在日本境内设计的幼儿园、保育园，以及其他儿童设施已超过 500 所。致力于传播以儿童为中心的环境与设计理念。

Coordinator of Hibino and its brand "Children's City" - Children's Facilities Project , chairman of kids design labo. Born in 1972 in Kanagawa, Japan. Graduated from Kogakuin University in Tokyo.
More than 500 kindergartens, nursery gardens and other children's facilities designed to date in Japan, dedicated to the dissemination of a child-centered environment and design philosophy.

恐龙
恐龙大世界

在旧北京城，孩子们常常在街上玩耍。但随着城市的发展，他们在街上玩耍的场景无法被看到。基于这样的背景，创建了这个概念，每个模块的放置都是为了让孩子们像街头一样玩耍。此外，这个规划对于这个场地的复杂形状是有效的，即半切椭圆，以及4米高的房间高度的骨架空间。每个模块都非常灵活，可以根据教育程序自由地打开和关闭墙壁。通过在上侧设置扶手，有一个像阁楼一样的空间，在那里可以让孩子们动态地学习和玩耍。像城市街道一样的每个过道都有空间供他们阅读书籍、聚会和玩耍。因此，可以创建这样的地方，让孩子们可以专注于他们的活动，并以他们的好奇心进行游戏和探索。关于颜色，只使用简单的两种，人们可以感到温暖。一个是模块的外墙，第二个是模块内的地板、墙壁和天花板。此外，模块的单元化对于相同材料的供应和维护也有好处。在这个独特的模块规划中，孩子们可以集中精力学习、玩耍和做任何事情。另一方面，街道连接每个模块，人们可以看到和认识彼此。

In old Beijing city, children used to play in street. But as the city developed, the scene of their playing in street could not be seen. Based on this background, this concept was created each module is placed as to make spaces for children to play just like in street. Also, this planning is effective for the complexed shape of this site, that is a half-sliced ellipse, and the skeleton space with 4m of room height. Each module is so flexible that walls can be put on and off freely according as the education program. By setting handrail on the upper side, there is a space like loft, where can be the place for children to play dynamically and learn. Each aisle like street in city has the space for them to read books, gather and play. As a result, the places could be created that children can concentrate on their activities and play and explore with their curiosity. About the color, only 2 kinds are used to be simple and people can feel warmness. One is in the exterior wall of module, and the second is the floor, wall and ceiling inside the module. Moreover, unitization of module makes a benefit on the supply of same materials and maintenance. In this unique planning with the module, children can concentrate on learning, playing, and doing anything. On the other hand, the street connects each module and people can see and recognize each other.

YM 托儿所
YM Nursery

项目地点：日本 / Location : Japan
项目面积：1145 平方米 /Area : 1145 m²
设计师：大沽・日比野拓 / Designer : Taku Hibino

IAI最佳人文关怀大奖 | IAI最佳设计大奖

这是日本鸟取县米子市的托儿所项目。这个地方被自然环境、海洋和山脉所环绕。然而，托儿所的以前环境是不变的，和一般的托儿一样，并不感性。尽管与大自然生活在一起，孩子们仍然无法感受到自然。因此，我们将其置于设计理念“感受自然，增强灵敏度”。让孩子们在日常生活中感受自然，我们专注于材料。每种材料都是有意义的，并且是为儿童选择的，可以通过以下三种方法看出。

This is a nursery project in Yonago, Tottori, Japan. This site is surrounded by the environment rich in nature, sea as well as mountain. However, the environment of the former nursery is unvarying and not sensuous, same as general nursery. Children couldn' t feel nature in spite of living with nature. Hence, we put it on the design concept, “Feel nature, grow sensitivity” . For children to feel nature in daily life, we focused on materials. Each material is meaningful and are chosen for children, which can be seen in the following three approaches.

首先，为了让孩子感受到更多的自然，我们使用了许多天然材料，但这不仅仅是出于这个原因。通过用手和脚触摸天然材料来刺激大脑的敏感，这将对他们的发育产生很好的影响。我们还想要促进儿童的敏感性和发展。每个木地板的树种、形状、布局和外观都不同，当然也可以用脚触摸。此外，在地板标志中，我们使用四种金属，每种材料都有触感。因此，孩子们可以触摸和感受多种天然材料，他们的敏感性将会增长，他们的想象力和大脑发育也会如此。

其次，当地的材料将给孩子们学习当地的机会。入口处的墙壁泥浆和建筑物周围的岩石部分来自当地。用于扶手的木材叫做 Chizu Cedar，也来自当地。此外，班级标志由 Yumihama Gasuri 制成，这里是传统的面料。每个班级的名字都是以山中的独特花朵命名的，这些花朵都盛开在 Daisen 这个站点附近的。随着孩子长大，他们的班级名字的花朵在山的更高点绽放。以这些花为主题，设计了班级标志。

First, to make children feel more nature, we use many natural materials. But it' s not only for this reason. The sensible stimulus taken by touching natural materials by hand and feet, stimulate their brain and it will have a good influence on their development. we also aimed to appeal to children' s sensitivity and development. Each wooden floor is different in tree species, shapes, lay-outing and looking, of course touching by foot. Additionally, in floor signs we use four kinds of metals. Each material has each sense of touching. Hence children can touch and feel many kinds of natural materials, their sensitivity will be grown, and their imagination and brain development as well.

Second, local materials will give children chances to learn their local. The mud of the wall at the entrance and the rocks in gabion surrounding the building are partly from the local area. The wood used for handrails is called Chizu Cedar which is also from the local. In addition, the class sign is made of Yumihama Gasuri, the traditional fabric here. Each of class name is named for the unique flowers which are bloomed in Mt. Daisen near this site. As children get older, the flowers of their class name are bloomed at the higher point in the mountain. With these flowers as motifs, the class signs are designed.

因此，孩子们可以通过许多天然材料获得很多机会在日常生活中了解当地人。

第三，为了讲述这个托儿所的历史，我们重新使用了一些旧建筑的材料。石笼中的岩石来自当地，其他岩石来自前托儿所拆除时的混凝土废弃物。课后日托中心建在托儿所旁边，其地板，墙壁和天花板由旧托儿所使用的地板材料制成。这些材料具有温暖，让人们回想起旧的风景。更重要的是，对于孩子来说，这将是一个了解我们如何妥善照顾事物的机会。

如上面提到的这三种方法，我们在不同的观点中使用了多种材料。每种材料都是天然的，而不是人造的，儿童可以感受到它的质地和故事，例如它们居住的环境和托儿所的历史。通过这些方式，孩子们在日常生活中玩耍时会增加敏感度。

Thus, children can get many chances to know about the local in their daily life with many natural materials.

Third, to tell the history of this nursery, we reused some materials from old buildings. Rocks in the gabion are some from the local, and the others are from concrete wastes at the demolition of the former nursery. And the after-school day care center is built next to the nursery, whose floor, wall and ceiling are made of floor materials used in the old nursery. These materials have a warmth and make people recall the old scenery. What is more, for children it will be a chance to learn how we take good care of things.

As referred in above these three approaches, we used many kinds of materials in different points of view. Each material is natural, not artificial and children can feel its texture and its story such as the surroundings where they live and the history of their nursery. In these ways, children will grow their sensitivity just while spending a daily life playing.

M. N托儿所
M. N Nursery

项目地点：日本 / Location : Japan
项目面积：393 平方米 / Area : 393 m²
设 计 师：大沽・日比野拓 / Designer : Taku Hibino

IAI 设计优胜奖

大沽·日比野拓 Taku Hibino

日比野公司及旗下品牌"幼儿之城"幼儿设施项目的统筹负责人，KIDS DESIGN LABO 董事长。1972 年出生于日本神奈川县，毕业于工学院大学。他截至目前在日本境内设计的幼儿园、保育园，以及其他儿童设施已超过 500 所。致力于传播以儿童为中心的环境与设计理念。

Coordinator of Hibino and its brand "Children's City" - Children's Facilities Project , chairman of kids design labo.
Born in 1972 in Kanagawa, Japan. Graduated from Kogakuin University in Tokyo.
More than 500 kindergartens, nursery gardens and other children's facilities designed to date in Japan, dedicated to the dissemination of a child-centered environment and design philosophy.

这是私人托儿所的室内装修项目，其主人是婚礼公司。普通的托儿所无法满足该公司员工的需求，因为他们不仅经常在工作日工作，还在假期工作。考虑到这种工作方式，该公司决定全年开放托儿所。在托儿所，这家婚礼公司提出了管理身份。这是一个想法，每个人都对这里的一切感到高兴。对于父母来说，他们可以享受抚养和工作。对于孩子们来说，他们可以很高兴地感受到自己的成长。而且，对于公司和当地社区来说，这个托儿所将有很好的影响力。为了让每个人都满意，这个托儿所的规划应该是灵活的，需要考虑用户和间接影响，如对当地社区的影响。

This is an interior renovation project of a private nursery whose owner is a wedding company. Common nursery couldn' t satisfy the needs of this company' s employees because they often work not only weekdays but also in holidays. Considering this situation of their working style, this company decided to make a nursery open all around year. In this nursery, this wedding company put a management identity. It is the idea that everyone can be happy at everything here. For parents, they can enjoy raising as well as working. For children, they can get glad to sense their own growing up. And also, company and local community, this nursery will have good influence. To make everyone happy, the planning of this nursery should be flexible and required consideration of users and the indirect influence such as a local community.

HN托儿所
HN Nursery

项目地点：日本 / Location : Japan
项目面积：2651 平方米 /Area : 2651 m²
设计师：大沽 · 日比野拓 / Designer : Taku Hibino

大沽 · 日比野拓
Taku Hibino

日比野公司及旗下品牌“幼儿之城”幼儿设施项目的统筹负责人，KIDS DESIGN LABO 董事长。1972 年出生于日本神奈川县，毕业于工学院大学。他截至目前在日本境内设计的幼儿园、保育园，以及其他儿童设施已超过 500 所。致力于传播以儿童为中心的环境与设计理念。

Coordinator of Hibino and its brand "Children's City" - Children's Facilities Project, chairman of kids design labo. Born in 1972 in Kanagawa, Japan. Graduated from Kogakuin University in Tokyo.
More than 500 kindergartens, nursery gardens and other children's facilities designed to date in Japan, dedicated to the dissemination of a child-centered environment and design philosophy.

这是神奈川县的一个幼儿园项目。这个幼儿园是由父母建立的，他们想要在丰富的自然环境中抚养孩子。为了满足他们的期望，通过利用这里的丰富自然资源，我们计划将这个幼儿园设计为儿童可以感受到大自然的地方。充满自然的建筑将让孩子们成长，发挥兴奋和刺激的机会，这将发展他们的感性和创造力。因此，将“整个日子感受到自然”作为这里的概念。一般来说，儿童在幼儿园室内玩现成的玩具或使用不灵活的材料。另一方面，当他们在外面玩耍时，有很多有趣的事情要发现，如花、草、树、昆虫等。这些自然元素根据天气和每个季节不断变化。它们必然鼓励孩子们比现成的玩具更具创造性和想象力。

This is a nursery project in Kanagawa. This nursery was founded by parents, who wanted to raise children in rich natural environment. To meet their expectation, by making use of rich nature surrounding here, we planned to design this nursery as the place where children can feel nature a lot. The building with full of nature will make chances for children to grow and play excited and stimulating, which will develop their sensibility and creativity. Hence, “Feel Nature the Whole Day” is put as the concept here. Generally, children play with ready-made toys or materials whose usage is inflexible, inside the nursery room. On the other hand, when they play outside, there are many interesting things to discover such as flower, grass, tree, insects, and so on. And these natural elements are changing continuously according to the weather and each season. They must encourage children to play more creative and imaginative than ready-made toys.

SMW 托儿所
SMW Nursery

项目地点：日本 / Location：Japan
项目面积：472 平方米 / Area：472 m²
设 计 师： 大沽・日比野拓 / Designer：Taku Hibino

大沽 · 日比野拓 Taku Hibino

日比野公司及旗下品牌"幼儿之城"幼儿设施项目的统筹负责人，KIDS DESIGN LABO 董事长。1972 年出生于日本神奈川县，毕业于工学院大学。他截至目前在日本境内设计的幼儿园、保育园，以及其他儿童设施已超过 500 所。致力于传播以儿童为中心的环境与设计理念。

Coordinator of Hibino and its brand "Children's City" - Children's Facilities Project，chairman of kids design labo.
Born in 1972 in Kanagawa, Japan. Graduated from Kogakuin University in Tokyo.
More than 500 kindergartens, nursery gardens and other children's facilities designed to date in Japan, dedicated to the dissemination of a child-centered environment and design philosophy.

这是一个托儿所，位于日本神奈川县Zama的低层住宅区。需要建立一个新的托儿所，以便减少不少于110名等待进入托儿所的儿童。在这个项目中，我们从以下背景中将"培育自治"作为概念。

最近，儿童游玩的环境发生了很大变化。离我们越近的城市，孩子们想象力就越少。即使在公园里，玩耍也受限于过于严格的规则。成年人认为孩子们独立地玩游戏非常好，这使儿童自己思考和行动的机会减少了，而课后被动学习的机会也增加了。从这些背景，我们设计了苗圃"培育自治"的4种方法。

This is a nursery, located at the part of a low-rise residential area in Zama, Kanagawa, Japan. It is required to build a new nursery so that no less than 110 children on waiting list to enter nurseries should be reduced. In this project, we put "Nurturing Autonomy" as the concept here from the following background.
Recently, the environment where children play has changed a lot. The closer to the city we go, the less place to play with our own bodies and imaginations there are. Even in the park, plays are limited by too strict rules to keep safe. It is hard to say that adults consider the children's independent play very well. The opportunities that children think and act by themselves are reduced, while the opportunities of passive learning in after-school has increased. From these backgrounds, we designed a nursery "Nurturing Autonomy" with 4 approaches.

瓦卡吉幼儿园
Wakagi Kindergarten

项目地点：日本 / Location : Japan
项目面积：920 平方米 /Area : 920 m²
设计师 / Designer : Gen Uchida

IAI 设计优胜奖

JAKUETS

Gen Uchida

我们公司设计游戏环境，创造未来价值。并致力于为儿童创造更高价值的环境设计。

Our company designs the play environment and creates future value.And aspires to create higher value environmental design for children.

这个项目是为了重建福岛县磐城的一所幼儿园。2011年地震发生后，许多人进出城市，人们与当地社区脱节是一个问题。客户希望幼儿园成为儿童和当地居民可以互动的地方，父母希望孩子通过与不同年龄的孩子交流来学习社交技能，希望当地居民保持旧的联系，所有这些都成了这个项目的基础。因此，儿童和当地居民的"互动"是我们的关键词，因为我们的目标是使这个幼儿园成为社区的核心。游乐场有两个层次，狭窄的空间，没有围墙的教室，在所有孩子一起吃饭的午餐室，我们添加了许多不同年龄的孩子可以在一起度过更多时间和交流的地方。在施工期间，我们举办了这些日子以来一直在减少的传统活动，以创造与当地居民沟通的机会。我们还设计了相邻的神社和桑多（从大门进入神社的方法）成为幼儿园的一部分，旨在使每个人都能使用和喜爱这个地方。最近，幼儿园往往被社区所关闭，但我们希望幼儿园与社区联系更紧密。

This project was to rebuild a kindergarten in Iwaki, Fukushima. After the earthquake in 2011, many people moved in and out of the city and people' s disconnection from the local community has been a problem. Client' s wish to make the kindergarten a place where the children and local residence could interact, parent' s expectation for their children to learn social skills by communicating with children of different ages, the wish of local residence to keep the old connection, all of these became the base of this project. Therefore "Interaction" of children and local residence was our keyword as we aimed to make this kindergarten the core of the community. Playground with two levels for a narrow space, classrooms with no walls, lunchroom where all children eat together, we added many places where children of different age can spend more time together and communicate. During the construction, we held traditional events that has been decreasing these days to create an opportunity to communicate with the local residence. We also designed so the adjacent Shrine and Sando (the approach to the shrine from the gate) to be part of the kindergarten and aimed to make the place being used and loved by everyone. Recently kindergartens tend to be closed from the community but we hope that the kindergarten will be more connected with the community.

二沙岛未来教育馆

Er Sha Island Future Education Hall

项目地点：中国广州 / Location : Guangzhou, China
项目面积：3500 平方米 / Area : 3500 m²
设计师：李友友 / Designer : Evans Lee

IAI 设计优胜奖

李友友 Evans Lee

李友友毕业于南大及中山大学 MBA 工商管理硕士，进修于中国美院现代设计高研班及亚太酒店设计高研班，于 2015 年开展个人事业，成立李友友室内设计工作室。

Founder Mr. Evans Lee graduated from NTU and Sun Yat-Sen University with an MBA in Business Administration. He studied at the China Academy of Art Modern Design High School and the Asia Pacific Hotel Design Institute. He started his career in 2015 and established Evans Lee Interiors Limited.

环观室内空间，大堂造型简约，显得落落大方，加上家具用色素雅，点染出宽敞的空间感，流露不凡气度。所有空间铺设纹理突出的木地板，多添一份自然气息，结合户外空间的绿树，使整体更加舒适写意。阳台经设计师重新修饰，把户外的鲜花丛林感觉搬入园区，亦成为小朋友小巧的休闲区，小朋友及老师们可从教室里悠然望景，享受那份写意和宁静。此外，设计师从细节处着手，为儿童卫生间增添丰盈质感，带来自然且富有层次的视觉效果，玩味十足。

Round view indoor space, lobby modelling is contracted, appear drop is easy, add furniture to use color simple and elegant, point dyeing gives capacious space feeling, reveal uncommon air. All space laid grain outstanding wood floor board, add a natural breath more, combine the green tree of outdoor space, make whole more comfortable freehand brushwork. Balcony by the designer to modify the outdoor flowers, jungle feel moved into the park, has become a child a small recreational area, children and teachers from the classroom carefree belvedere, enjoy the freehand brushwork and peace. In addition, stylist begins from detail place, add rich simple sense for children toilet, bring natural and rich visual effect of different levels, playfulness is dye-in-the-wood.

Conceptual Design

室内 —— 方案设计

信达泰禾
Shinta Taishe Shanghai Courtyard

项目地点:中国上海 / Location : Shanghai, China
项目面积 : 3600 平方米 / Area : 3600 m²
公司名称 : 上海上榀建设工程有限公司 / Organization Name : GP Design Constrution

IAI 设计优胜奖

上榀设计 GP DESIGN

上榀设计成立于1998年，专业服务于地产商的售楼会所、样板间、酒店、商业等室内空间的设计。近年来的代表作品有泰禾集团的上海院子、杭州院子等。合作伙伴包括全球以及国内的知名开发商，如：泰禾、金茂、阳光城、卓越、华润、中海、万科、龙湖、光明等，设计业务至今已覆盖全国二十多个大中城市，作品屡获业内专业奖项认可。

Founded in 1998, GP DESIGN specialize in the design of interior spaces for real estate developers, such as sales clubs, showrooms, hotels, and commercial buildings. Representative works in recent years include the Shanghai Yard, and the Hangzhou Yard.GP DESIGN partners include well-known developers around the world, such as: Tahoe, Jinmao, Yango, Excellence, CR Land, China Overseas Property, Vanke, Longhu, Chiway Land, Bright Life, Sansheng Hongye, etc. The design business has covered the whole country. More than 20 large and medium-sized cities have won numerous industry awards.

上榀设计凭着优良精湛的设计水准在业内赢得了良好的口碑，并矢志不渝、用心钻研。此次与泰禾集团合作的上海院子项目设计采用新中式设计理念，空间绚丽古风，开启一场扣人心弦的逐美之旅。审美力与鉴赏力并驾齐驱，驭艺术之张弛，铸就空间骄傲的力量。流淌于血脉的文化情节在此升华、喷薄出华贵与灵动共生的非凡气质。

GP DESIGN has won a good reputation in the industry with its excellent and exquisite design standards, and is determined to study hard. The design of the Shanghai Yard project in cooperation with Taihe Group adopts the new Chinese design concept, and the space is gorgeous antique, opening a journey of exciting beauty. The aesthetic power and the appreciation power go hand in hand, and the relaxation of art creates the power of space pride. The cultural plot that lingers in the blood is sublimated here, and the extraordinary temperament of luxury and symbiosis is sprinkled.

卓越地产
Eccellence Real Estate

项目地点：中国上海 / Location : Shanghai, China
项目面积：103000 平方米 / Area : 103000 m²
公司名称：上海上榀建设工程有限公司 / Organization Name : GP Design Constrution

IAI 设计优胜奖

上榀设计 GP DESIGN

上榀设计成立于 1998 年，专业服务于地产商的售楼会所、样板间、酒店、商业等室内空间的设计。近年来的代表作品有泰禾集团的上海院子、杭州院子等。合作伙伴包括全球以及国内的知名开发商，如：泰禾、金茂、阳光城、卓越、华润、中海、万科、龙湖、光明等，设计业务至今已覆盖全国二十多个大中城市，作品屡获业内专业奖项认可。

Founded in 1998, GP DESIGN specialize in the design of interior spaces for real estate developers, such as sales clubs, showrooms, hotels, and commercial buildings. Representative works in recent years include the Shanghai Yard, and the Hangzhou Yard. GP DESIGN partners include well-known developers around the world, such as: Tahoe, Jinmao, Yango, Excellence, CR Land, China Overseas Property, Vanke, Longhu, Chiway Land, Bright Life, Sansheng Hongye, etc. The design business has covered the whole country. More than 20 large and medium-sized cities have won numerous industry awards.

上榀设计在中国的许多城市完成了一系列具有规模代表性的设计作品。此次与卓越集团合作的卓越地产项目，空间注重功能的同时强调色彩、材质、尺寸、细节及饰品等给人的感观体验，打造出气质不凡、庄重沉稳的高品质空间。

GP DESIGN has completed a series of representative design works in many cities in China. This outstanding real estate project with the Excellence Group focuses on the function while emphasizing the sensory experience of color, material, size, detail and accessories, creating a high-quality space with extraordinary temperament and solemnity.

阳光城檀院
Sunshine City Tan Yuan

项目地点：中国浙江 / Location : Zhejiang, China
项目面积：1993 平方米 / Area : 1993 m²
公司名称：上海上榀建设工程有限公司 / Organization Name : GP Design Construction

IAI 设计优胜奖

上榀设计与阳光城集团合作的檀院项目设计采用新中式设计理念，空间注重功能的同时强调色彩、材质、尺寸、细节及饰品等给人的感观体验，体现了典藏艺术情怀，沉醉东方韵味的主题。

The design of the Tanyuan project in cooperation with the Sunshine City Group adopts the new Chinese design concept. The space emphasizes the function while emphasizing the sensory experience of color, material, size, detail and accessories, reflecting the artistic feelings of the collection and intoxicating the oriental charm. Theme of.

黑暗
Dark

项目地点：中国湖南 / Location : Hu nan, China
项目面积：147 平方米 /Area : 147 m^2
设 计 师：陈呈 / Designer : Chen Cheng

IAI 设计优胜奖

陈呈 Chen Cheng

2018 第三届包豪斯国际设计大赛铜奖
40 under 40 中国（湖南）设计杰出青年(2018-2019)
2018 IAI 全球设计大奖优胜奖
2018 法国双面神设计大奖
2018 湖南顶峰设计奖铜奖

THE THIRD BAUHAUS AWARD INTERNATIONAL DESIGN COMPETITION
40 under40 China Hunan Design Outstanding Youth
2018 IAI Design Award Winner
FRENCH GPDP INTERNATIONAL DESIGN AWARD (2018-2019)
HUNAN APEX DESIGN AWARDS BRONZE AWARD (2018-2019)

客餐厅自由、开阔的空间功能布局，延续黑白对比的设计手法，用不同的黑色材质去演绎空间的质感，大胆地植入色彩，灯具有节奏地加入，强调非常规性的空间体验。

主卧在规划上因户型本身的结构特点，综合业主的生活方式和功能需求，做了少许调整，尽可能在空间形式上保持整体统一的同时，满足各种生活的需求。

设计语言的流动性和连续性，使玄关、客餐厅、卧室、卫生间结合成一个和谐的整体空间。

The liberal and open living room and dining room follow the black-white contrast design technique; different black materials are used to interpret the texture of the space, and the bold colors and lamps emphasize an unconventional experience of the space.
Considering the structural characteristics of the house layout, a few adjustments are made to the master bedroom plan to integrate the owners' life style and functional requirements and maintain consistency in spacial form while satisfying the various needs of life.
The fluidity and continuity of the design language allows the vestibule, living room and dining room, bedrooms and bathroom to make an overall harmonious space.

X别墅
X-Villa

项目地点：中国湖南 / Location : Hunan, China
项目面积：450 平方米 / Area : 450 m^2
设 计 师：陈呈 /Designer : Chen Cheng

IAI 设计优胜奖

入户玄关深灰色的地面材质、极简的木质楼梯扶手与层次丰富的吊顶在形式、材质上的对比，构成空间上的视觉张力和气场，开启这个空间的第一印象。

客餐厅的空间设计削弱电视媒体等娱乐功能，强调家庭生活最核心的功能：聊天、聚餐、品茶、看书、育儿、增加家庭成员互动。古典的线条设计和现代家具的平衡，整体灰色系和局部艳色的平衡，营造优雅、轻松的空间气氛，给当下快节奏的生活带来心灵上的宁静。

The contrast of dark gray ground material and simple dark wooden staircase handrails with abundant ceiling in form and material constitutes the visual tension and air field in space, which opens the first impression of this space.

The space design of guest restaurant weakens the entertainment functions such as TV media and emphasizes the core functions of family life, such as chat, dinner, tea, reading, parenting, and increasing family members'interaction. The balance of classical line design and modern furniture, the balance of overall grey system and local bright color, create elegant and relaxed space atmosphere, bring spiritual tranquility to the current fast-paced life.

天荟悦麓

Todtown Tianhui Yuelu Elite Housing

项目地点：中国上海 / Location : Hunan, China
项目面积：186 平方米 / Area :186 m²
公司名称：邸设空间设计有限公司 / Organization Name : Decor&Decor Interior Design(Shanghai)

IAI 设计优胜奖

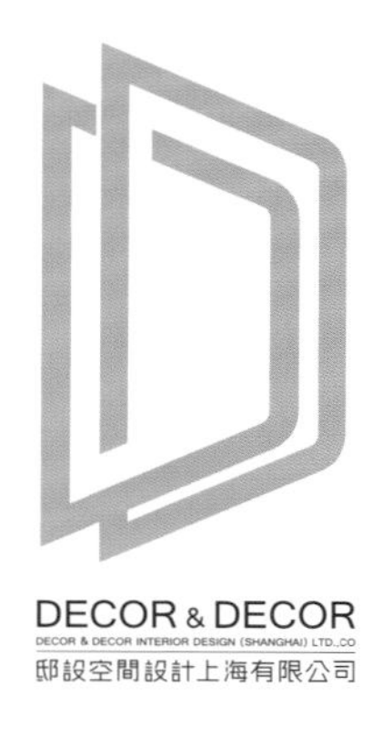

邸设设计 Decor & Decor Interior Design (Shanghai)

成立于 2015 年，由从业十几年的国际国内软装主创共同打造的国际设计团队，设计师主要是业内顶尖女性设计师，主打室内硬装、软装设计与产品改造，品牌秉承着对精致生活的追求，打造室内软装设计结合家居产品私人定制设计为一体的软装设计模式。为居住者呈现更舒适、更精致的室内设计同时，融入居住者个人性格与形象气质的私人定制空间。

DiShe（Décor & Décor）Interior DesignCo., Ltd was founded in 2015, the international design team created by the international and domestic soft-wearing designers who have been engaged in the industry for more than ten years, the designers are mainly the top female designers in the industry, focusing on the interior hard-wearing, soft-packing design and product transform ation. The brand adheres to the pursuit of exquisite life and builds The soft design of the interior is combined with the custom design of the home product. For the occupants to present a more comfortable, more refined interior design, while incorporating the personal custom of the occupants' personal character and image.

TODTOWN天荟悦麓项目坐落于上海莘庄地铁站的绝佳地段，都市的繁华、健康的调理、人文的熏陶和交通的便利，让这里成为一个富有前瞻性的强大枢纽中心。在这里，俯瞰繁华，浮华褪去，回归最本心的世界。

天荟悦麓项目花园住宅，无论是从统摄全局还是引领目标人群的重要定位上，邸设空间设计特别关注消费者的需求特征：舒适，在满足居住功能的同时，享受精神世界的充实。基于这两点，设计者的理念是：美丽、奢华与名望。

TODTOWN Tianhui Yuelu Project is located in the excellent district of Shanghai Xinzhuang Metro Station. Owing to urban prosperity, healthy conditioning, humanistic edification and traffic convenience, it becomes a strong hub center with great perspectiveness. Overlooking prosperity and taking off pomposity, people can return to the world of conscience.

The garden residence of Tianhui Yuelu project pays more attention to consumer demand features in residence space design, as governing the overall situation or guiding target population: comfort, which makes people enjoy enrichment of spiritual world, while satisfying dwelling functions. Based on it, the philosophy of designers is targeted at beauty, luxury and good reputation.

山月间
Intermountain Homestay

项目地点：中国广东 / Location : Guangdong, China
项目面积：246 平方米 /Area : 246 m²
设 计 师：葛凡 / Designer : Ge Fan

IAI 设计优胜奖

葛凡 Ge Fan

中国建筑装饰协会高级室内建筑师
珠海市平面设计协会执行会长
珠海市动漫协会副会长
设计理念：凡而有务，物而非凡，对空间设计有着纯粹的追求，希望能够一直坚持自己，创造出真实、有感染力的作品。

China Building Decoration Association
Senior Interior Architect
Executive Director of Zhuhai Graphic All that have affair, contaent and especially
Vice President of Zhuhai Animation Association
Design concept:
All that have affair, content and especially, There is a pure pursuit of space design, hoping to always adhere to oneself and create a real and infectious work.

房屋为土夯坭墙、砖、瓦木结构，是典型的岭南乡村民居建筑。我们在不严重破坏建筑的基础原貌下对房屋进行设计改造，并且配合项目的设计需求，在尊重排山村及房屋的历史文化前提下，合理保留、融入相关人文元素。发扬排山村的质朴文化，传承传统工匠精神与纵情于乡村山野情怀，提高当地传统文化自信心。

在设计构思上，头门造型主打新中式解构，强调岭南文化，合理结合传统建筑元素；建筑的大部分夯土墙面基本保留，保存当地原有的建筑文化特色，庭院新建墙面可使用普通砖、玻璃砖以及金属板相结合，做到现代与传统材质在墙面上的融合。

设计上使用的固件、装饰品尽量使用或改造有年代感而经典的老旧物件，如门板、琉璃、砖雕、老瓦片、石狮等。

The houses are earthen walls, bricks and tile structures. They are typical Lingnan rural residential buildings. We will design and renovate the house without seriously damaging the original structure of the building, and in accordance with the design needs of the project, we will reasonably retain and integrate relevant human elements under the premise of respecting the historical culture of the villages and houses. Carry forward the simple culture of Paishan Village, inherit the spirit of traditional craftsmen and indulge in the feelings of rural mountains and rivers, and enhance the confidence of local traditional culture.
In terms of design concept, the headgate model styling focuses on the new Chinese structure, emphasizing Lingnan culture and rationally combining traditional architectural elements; most of the building's earthen walls are basically preserved, and the original architectural and cultural features are preserved. The combination of common brick, glass brick and metal plate makes the fusion of modern and traditional materials on the wall.
The firmware and decorations used in the design should be used as much as possible or retrofitted with old and classic objects such as door panels, glazes, bricks, old tiles, and stone lions.

弗洛斯秘境
Forest Secret Restaurant

项目地点：中国台湾 / Location : Taiwan, China
项目面积：1985 平方米 /Area : 1985 m²
公司名称：大言室内装修有限公司 /Organization Name :
设 计 师：黄金旭 /Designer : Huang Jinxu

黄金旭 Huang Jinxu

大言室内装修有限公司设计总监
中原大学室内设计研究所硕士
逢甲大学建筑专业学院兼任副教授

Principal Designer of Great Word Design Company;
Master of Institute of Interior Design, Chung Yuan University;
Associate Professor, School of Architecture, Feng Jia University

在空间场域上以森林的生态意象形塑了“绿—生长”与“红—蜕变”两个用餐区。在“绿—生长”区以大片紫藤色干燥花、几何图形地毯铺面、动物系的彩绘墙面，以及米灰色与墨绿的沙发错落摆放，为空间注入生机。而“红—蜕变”区则选用椰褐色的干燥花，及绣蚀感的特殊墙面来呼应着纱质叶脉隔屏，料理台的黑色大理石及酒红色餐椅的连接，让蜕变后的优雅与别致在空间扩展延伸。设计师巧妙地运用不同的灯光色调及阴影变化，加强营造了两个用餐环境的差异性。

The dining area is divided into two zones, namely “Green: Growth” and “Red: Transformation.” Both are designed with the ecological imagery of forests. The “Green: Growth” zone contains walls with colorful animal paintings and geometrically patterned carpets decorated with dried flowers in the wisteria color. In addition, beige-grey and blackish green sofas have been haphazardly placed in both zones to help create a lively space. The “Red: Transformation” zone features dried flowers in the coconut brown colo, as well as walls with a rusty metal texture that mesh with gauze leaf-vein screens. The combination of black marble and wine-colored dining chairs invokes senses of elegance and delicacy, which extend across the space after the “transformation.” The designer aptly used different color tones of lighting, as well as shadowing changes, to create a contrast between the dining environments of the two zones.

九月森林样板房
Iforest Model House in September

项目地点：中国南京 / Location : Nanjing, China
项目面积：约 472 平方米 /Area :about 472 m²
公司名称：锦壹（上海）装饰设计有限公司 / Organization Name : Kim International Design Group
设 计 师：黄锦华 /Designer : Huang Jinhua

IAI 设计优胜奖

锦壹国际设计顾问机构
Kim International Design Group

KIM 锦壹国际设计是中国陈设艺术行业标杆品牌机构，致力于商业空间、精品酒店、店铺企划营造一体化代建、情景体验式售楼处、样板房、及高端住宅的高定设计，从空间美学设计出发到生活美学的创造传播，为中国专业的美好生活场景营造商。

KIM International Design Group limited is a benchmarking brand organization in China's furnishings art industry. It is dedicated to the commercial space, boutique hotels, store planning and creation of integrated construction, scene-experienced sales offices, model houses, and high-end residential design. The aesthetic design starts from the creation and communication of life aesthetics, and creates a good life scene for Chinese professionals.

在某一个温暖的午后，拘一缕阳光，品一杯红茶，再捧上一本好书，慵懒地陷入沙发的温柔中，遐想一场跨越时空的对话，品读每一个字的浪漫，每一句话的温柔，惬意，缱绻，任时光从指缝溜走。

本方案，在深入了解项目定位后，结合现有外部环境和内部环境的统一，并接轨国际现下的生活方式营造给业主专属感、尊崇感、安全感、可参与感。如果说从千万色彩中选择一款最能打动人心的颜色，有着冷艳色泽质感的蓝调必属其一，从一丝柔和的清爽到一抹深沉的典雅，从一份沁透心脾的舒怡到一丝平静深邃的诱惑。

On a warm afternoon, a ray of sunshine, a cup of black tea, and then in a good book, lazily in the tender of sofa, daydream a conversation across space and time, read every word of romance,Every word is gentle, pleasant and deeply attached, Let time slip through my fingers.

This scheme, after thorough understanding project positioning, combined with the unity of the existing external environment and internal environment, and is in line with the international current way of life to build to the owner exclusive feeling, exalted feeling, a sense of security, can be engaged. If from the thousands of colors to choose the color of a most can move the heart, with elegant color texture blues must belong to the first, from a gentle and relaxed to wipe a deep and elegant, from a comfortable and pleasant seeps through mist to a quiet depth of temptation.

金基N2样板房
JIN JI N2 Model House

项目地点：中国南京 / Location : Nanjing, China
项目面积：472 平方米 /Area : 472 m²
公司名称：锦壹（上海）装饰设计有限公司 / Organization Name : Kim International Design Group

IAI 设计优胜奖

法式优雅属于高尚的格调和内敛的情怀，而这植根于法国人骨子里的罗曼蒂克，任何表象的花哨与浮华都无从表达。在陈设设计中，KIM结合淡雅的空间色调、金色几何线条与典雅时尚的床品，打造出一个法式轻奢主义的空间格调，既承袭了最具法国特色的优雅、浪漫，又让空间更为轻盈、时尚与现代，从而贴近时下青年对品质生活与潮流新风尚的理解。

在灵活的配饰选择上，则有更强的法式浪漫与创造性，强化主题风格的同时提升空间的艺术感。

French elegance belongs to noble style and introverted feeling, which is rooted in the romance of the French people, and no appearance of flowery and gaudy can be expressed. In display design, KIM, combined with the space of quietly elegant tonal, golden geometric lines and elegant fashion bed is tasted, create a French "style of luxury, both took their cue from the elegance and romance of the French features, at the same time make a space more lightsome, fashionable and modern, and close to the current youth understanding of quality life and fashion trend.

In the flexible accessory choice, there is stronger French romance and creativity, strengthening the theme style while at the same time intended to enhance the artistic sense of space.

燕西书院90合院
Yanxi Academy-90 Courtyard Model House

项目地点：中国北京 / Location : Beijing, China
项目面积：283 平方米 / Area : 283 m²
设 计 师：黄婷婷、杨俊 /Designer : Huang Tingting, Yang Jun

IAI 设计优胜奖

黄婷婷　Huang Tingting
杨俊　Yang Jun

天鼓设计创始人及设计总监，致力于室内设计行业二十年，具有建筑及结构学术背景，擅长从建筑出发，结合景观，做到建筑与室内一体化设计。追求完美精品，深得地产甲方肯定，并获得多项室内设计大奖。

As co-founder and chief designer of Shanghai Tian Gu Interior Decoration Design Co. Ltd, Mr. Yang has dedicated himself to interior design for 2 decades. Being an expert in building industry and structural engineering, he is skilled in extraordinary integration of architecture and interior design. Pursuing perfection has always been his goal. His works are constantly appreciated by a great many real estate developers. He has won a number of interior design awards.

沉醉是一件美好的事情。“看山是山，看水是水，看山不是山，看水不是水，看山还是山，看水还是水。”只有沉醉才能体悟不同的人生境界。我们试图通过安静有序的空间设计，使居者沉醉于此，更安于起居、休闲、思考、放松。

主人、老人、孩子，三代人各自私属空间，设计采用半围合与高低错落的屏风式手绘背景，加强空间层次，表达不同的心属故事，展现出祥和、恬美和静思的意念。

Indulging ourselves is beautiful and intriguing. Self-love means caring for oneself, setting boundaries for oneself, being truthful to oneself and trusting oneself. Self-love requires tremendous courage - having the strength to be who you are in the world that is desperately trying to change you and in return the world will love you. What we have managed to do is to provide a refined internal and external design, which gives you space to love yourself. It has been proven time and again, if you cannot love yourself, you will never be capable of loving another.
You, your parents and your kids, three generations, all have their personal space. The semi-enclosed design of scattered high and low artistic hand-painted and screen like backgrounds is an expressive way to enrich the spatial level and celebrate glorious heritage, culture and traditions. They are meant to rejoice special moments and emotions in our lives with our loved ones.

凤凰天誉华府
Phoenix is known to Washington

项目地点：中国深圳 / Location : Shenzhen, China
项目面积：100 平方米 / Area : 100 m²
公司名称：深圳市华贝软装饰设计有限公司 / Organization Name : Huabei Soft Adornment

IAI 设计优胜奖

华贝软装
Huabei Soft Adornment

深圳市华贝软装饰设计有限公司（HBD）由著名室内设计师靳冬先生于2011年创立，总部位于中国深圳，并在香港、西安、保定、郴州等地设有分公司。HBD专业从事样板间、售楼处、别墅、豪宅、酒店、办公空间室内设计及艺术品开发服务，赢得客户和社会的一致认可和高度赞赏，为顶级品牌室内设计机构之一。

Shenzhen Huabei Soft Decoration Design Co., Ltd. (HBD) was founded in 2011 by the famous interior designer Jin Dong. It is headquartered in Shenzhen, China and has branches in Hong Kong, Xi'an, Baoding and Zhangzhou. HBD specializes in interior design and art development services for model rooms, sales offices, villas, luxury houses, hotels, office spaces, and has won unanimous recognition and high appreciation from customers and the society. It is one of the top brand interior design organizations.

餐厅夹丝玻璃墙面、背景墙都展示其材料的创新性，与餐具、餐桌和酒柜的金属色彩介入，构成点线面的艺术性，浑然天成。

主卧以曲线、直线形成独特的韵律感，金色的几何饰品以面的形式为空间制造戏剧性。在水晶灯流光溢彩的光芒下，奏响了空间的优雅独特的人文气质。

大自然的蓝绿色彩成了婴儿室的主色调，富有质感的墙纸以点线成面，搭配纯天然棉质床品，跨越居住空间的界限，突显人文与艺术的关系。

The wired glass wall and the background wall show the innovation of its materials, and the intervening of the metal color of the tableware, dining table and wine cabinet, constitutes the artistic aspect of the point, line and plane, like nature itself.
The master bedroom forms a unique rhythm with curves and straight lines, and the golden geometric ornaments create a dramatic space for the space in the form of faces. Under the radianc of flowing light and color of crystal lamps, the elegant and unique humanistic temperament of the space is played.
Natural blue-green color has become the main color of the baby room, and the textured wallpaper is plane with lines and dots and natural cotton bedding, which crosses the boundaries of living space and highlights the relationship between humanity and art.

花间堂
Flower Memory

项目地点：中国贵州 / Location : Guizhou, China
项目面积：480 平方米 /Area : 480 m²
设 计 师：李楠 / Designer : Li Nan

李楠　Li Nan

东方卫视《梦想改造家》特邀设计师
上海李和杜筑作室内装饰设计有限公司创始人
上海海华设计艺术总监
“三联社”木兰会副理事长
伦敦艺术大学切尔西学院室内与空间设计硕士

DRAGON TV, Dream Home Designer
Shanghai Li & Du Zhuzuo Interior Decoration Design Co., Ltd. Founder
Haihua Design, Shanghai Art Director
Mulan Society of "Sanlian Design" Vice Chairman
MA Interior & Spatial Design, University of the Arts London, Chelsea

地下一层整层为老人层，给老人一个世外桃源，卧室、起居、中式餐厅和厨房、满院花草。流水亭台，甚至还有亲手打理的菜园，让他们的晚年生活丰富而有情调。一、二层为男女主人以及孩子的生活空间，一层囊括起居室、书房、客厅以及一个西式厨房，二层主要是卧室层。地下一层在需要时可以将老人的生活起居空间以及户外空间分隔开，配备的烧烤区、户外就餐区非常适合男女主人举办派对活动，从而充分利用每一个空间功能。

本案整体打造的是新中式风格，同时又考虑到不同年代人群的审美需求，地下一层整体氛围更偏向于传统中式，而在一、二层引入大东方的概念，提取中式韵味，搭配国际化的家具饰品，让整个空间显得简约时尚，又不乏传统美学的沉淀。

The whole floor of the basement is the old man's floor, giving the elderly a paradise, bedroom, living, Chinese restaurant and kitchen, full house flowers and plants, flowing water pavilions, and even a hand-washed vegetable garden, so that their old age life is rich and sentimental. The first and second floors are the living space for male and female owners and children. The first floor covers the living room, the study room, the living room and a western kitchen. The second floor is mainly the bedroom floor. The basement level separates the living space and outdoor space of the elderly when needed. The barbecue area and outdoor dining area are ideal for hostesses to take advantage of every space function
The designer is the dreamer, the wedding design is to create a dream of love, and the interior design is to create a dream of life. This case not only conveys the hostess's aesthetic concept, but also gives her client a dream of love and gives her family a dream of life.

罗绮香
Luo Qi Xiang

项目地点：中国贵州 / Location : Guizhou, China
项目面积：310 平方米 /Area : 310 m²
设 计 师：李楠 / Designer : Li Nan

IAI 设计优胜奖

原始建筑在布局上是没有一层的工作室部分，考虑到建筑自身一层使用面积小、使用率差的特点以及综合考虑设定的女主的职业特点，我们在一层增加了一个工作室，使一层变为一个工作、会客的功能空间。负一层则为家庭内部人员使用，形成一个很好的内外有别的空间布局。

在一层的走廊尽头，设置了一个挑空的观景台，更好地实现了一层跟负一层的互动。空间布局上为了让阳光更好地进入到每个空间，把墙体进行了合理的拆建，让功能之间进行更好互动。

In the layout of the original building, there is no one-storey studio part. Considering the characteristics of small floor area, poor utilization rate and the occupational characteristics of the female princess, we have added a studio on the first floor to make the first floor become a functional space for work and visitors, and the second floor is used by family members, forming a good spatial layout Other spatial layout.

At the end of the corridor on the first floor, an empty viewing platform is set up to better realize the interaction between the first floor and the negative one. In space layout, in order to let sunshine enter each space better, the wall is demolished and constructed reasonably, so that the function and function interact well.

无关西东
Nothing Related

项目地点：中国广东 / Location : Guangdong, China
项目面积：158 平方米 / Area : 158 m²
设 计 师：刘国海 / Designer : Liu Guohai

IAI 设计优胜奖

刘国海 Liu Guohai

毕业于华南理工大学，2001 年开始从事建筑及室内设计行业，2014 年作创办刘国海建筑设计事务所至今。在不同设计领域中不断获得尝试和实践，在建筑、园林、vi 系统及软装陈列设计里也多有涉猎，同时担任多家地产公司及材料设计公司的顾问工作。

He graduated from South China University of Technology and started to work in the architectural and interior design industry in 2001. In 2014, he founded Liu Guohai Architects. He has been continuously experimenting and practicing in different design fields. He has also been involved in architecture, garden, vi system and soft-pack display design, and has served as a consultant for many real estate companies and material design companies.

在这个项目里，设计师不想过多重复传统符号，也无意打坐"入禅"。在现代的生活中，用西式的"泊"来物件，"零乱"地散漫成一幅别样的东方画面。大面积的留白和朴素的用料体现出节制；文人的清高用简练、轻盈的线条勾勒，由这些线，串联彼此分离、独立的关系。没有一组组、一套套的界线来圈出轮廓，而是任其散漫存在。设计师也不希望用家具来定义风格，陈列和家具用了各种不一样的风格，从现代西方的，到典型的中式柜子。这种混和基于某种共性，而这个共性就是其内在淡淡的东方构成的神髓，包容的混搭由此也能传达闽粤市井表象下的文化包容性。同时以此体现不拘形式的生活态度。在样板间商业需求的前提下，尽量保留对节制和真实的主张。

In this project, we don't want to repeat too many traditional symbols, and either we don't want to Zen. We want to present the Modern life into a different kind of oriental picture by "disorderly" make up with Western-style objects. The large area of ?white space and plain materials reflect the restraint; We use the simple and lithe line to represent the aloof literati, and with all these lines connect the separated independent aspect. There is no boundaries but just let it exist. We also don't want to use furniture to define styles, so we use a variety of different styles, modern Western and typical Chinese cabinets. This kind of mixing is based on a certain commonality, which make up the oriental. The mash-up can also convey the inclusive cultural of the folk culture of Fujian and Guangdong city. At the same time, it reflects the casual attitude of life. Under the premise of commercial needs, we try to retain the idea of ??restraint and truth.

不二
Consistent

项目地点：中国江苏 / Location : Jiangsu, China
项目面积：150 平方米 / Area : 150 m²
设 计 师：童一宸 / Designer : Tong Yichen

IAI 设计优胜奖

童一宸 Tong Yichen

6年从业经验，2012年毕业于南京师范大学环境艺术设计专业，擅长现代、简欧、北欧、新中式风格。
设计理念：最好的设计，就是为客户体现其生活的态度，为空间还原其相宜的状态。

Six years of experience, graduated from Nanjing Normal University in 2012, Majoring in environmental art design, specializing in modern, simple European, Nordic, and new Chinese style.
Design concept: The best design is to reflect the attitude of the customer for the life and restore the appropriate state for the space.

本案布局上通过采用平实而不花哨的手法，在满足实地各类施工要求与实用需求的情况下，创造了大面积可变化、互动的空间感觉。造型以空间功能为主，顺结构而起，采用多种不同质感的元素，旨在以差异化的材质相互拼搭配合。如使用多种规格、纹理细腻的大理石砖拼配鱼骨拼木纹砖，与墙面一体延伸的设计手法配上不规则分割的大面积木饰面，使整个空间立面造型与底面、顶面相互区隔又相互统一。主卧更是以质感细腻、风格粗犷的纳米水泥与烤漆面柜体相碰撞，辅以木质电视柜与不锈钢开放柜的穿插，以不拘一格的手法突出空间的质感与品位。

On the layout of this case, by adopting the technique of plain and not fancy, the space feeling of large area can be changed and interactive is created under the condition of satisfying all kinds of construction requirements and practical demands.
The modeling takes the space function as the main function, along with the structure, uses many different texture elements, aims to match with each other by the difference material. Such as the use of a variety of specifications, fine texture of marble brick with fishbone wood tile, and wall extension of the design with a large area of irregular segmentation of wood decorative surface, so that the entire space elevation modeling and bottom, The top surfaces are separated from each other and united with each other. The main bedroom is characterized by exquisite texture, rough style of nano-cement and lacquer cabinet, wooden TV cabinet and stainless steel open cabinet.

尺度
Yardstick

项目地点：中国江苏 / Location : Jiangsu, China
项目面积：150 平方米 / Area : 150 m²
设 计 师：童一宸 / Designer : Tong Yichen

IAI 设计优胜奖

尺度，在生态学上是指准绳、分寸、衡量长度的定制，可引申为看待事物的一种标准。以此为解，它是个冰冷的词汇，与家的定义大相径庭，但对于我们而言，它是贯穿一切的准绳，是营造家、使家舒适的灵魂。家，就是一个不需要华丽的地方，一个不需要多大的地方。它应该是刚刚好的，像一件量身订制的衣服一样，不大也不小，不冷也不热。它就是"尺度"，是问题的解法，也是问题的答案，只以业主的需求出发，不媚不俗，不需要过多的言语，不需要多余的眼神，它会向所有来这里的客人介绍，这就是业主，这就是业主的家，一个一切都刚刚好的归处。

In ecology, it refers to the customization of standard, measure, and measure length, which can be extended to a standard of looking at things. For this reason, it is a cold word that is quite different from the definition of home, but for us it is the guiding principle through which to build a home and make it comfortable.
Home is a place that doesn't need to be gorgeous, a place that doesn't need much. It should be just right, like a tailored clothes, not quite not small, Moderate temperature.It is the "yardstick", the solution to the problem, and the answer to the problem. It is only based on the needs of the owner. It is not vulgar, it does not need too much speech, and it does not need extra eyes. It will introduce to all the guests who come here, this is the owner. This is the owner's home, a place where everything is just right.

太原盛高
Taiyuan Andaluz Manor 150 Type

项目地点：中国山西/ Location : Shanxi, China
项目面积：150 平方米 / Area : 150 m²
设 计 师：王黎贤 / Designer : Wang Lixian

IAI 设计优胜奖

法式家居以其特有的内敛特质与脉脉温情，传达着设计师对设计环境的诗意追求。

避免了法式繁复雕刻元素的堆砌，设计师以更为简洁、克制的家具造型装饰空间。色彩上遵循法式回归自然的特点，使整体软装低调温柔而不至于平淡。

French style residential design which with unique restrained characteristics and tenderness , conveys the poetic pursuit of the designer for the design environment.
To avoid piling up the French intricate carved elements, the designer decorates with the furniture which is more concise and restrained. The colors of the whole space follows the French characteristics that is to return to nature , and so that the overall soft is understated, gentle and not dull.

王黎贤 Wang Lixian

益善堂装饰设计创始人，1974 年出生于江苏省无锡市，毕业于常熟理工大学工艺美术专业，于上海交通大学进修 MBA 学位。有着丰富设计背景的他偏爱欧洲文化，追求美式古典风，不强调繁复的雕刻，以舒适和多功能为主，强调简洁、明晰的线条和优雅、得体有度的装饰，但是也不失高雅和尊贵的感觉。

Yishantang decoration design founder, born in 1974 in Wuxi City, Jiangsu Province, graduated from Changshu University of Technology, majoring in arts and crafts, and engaged in advanced studies Shanghai Jiaotong as MBA. With a rich design background, he prefers European culture, pursues American classical style, does not emphasize complicated carvings, and focuses on comfort and versatility, emphasizing simple and clear lines and elegant and graceful decoration, but also elegant and elegant. Distinguished feeling.

感知线构
Perception. Line Composition

项目地点：中国台湾 / Location : Taiwan, China
项目面积：251 平方米 / Area : 251 m²
公司名称：一研设计制研院 / Organization Name : One Researvch Design

IAI 设计优胜奖

一 研 設 計
ONE Research Design

一研设计
ONE Research Design

设计，是一种为了生活更好而衍生的行为。
这美好的普世价值，也是一研设计所追求的典范。
环境因地制宜，
空间因人设事。
以自然为师、与人文为伍。
为一研设计的精神。

Design is a kind of behavior that is derived for better life.
This wonderful universal value is also a model pursued by One Research design.
Environment . Space . Design are the main purposes of people.
The spirit of One Research design is to take nature as a teacher and associate with the humanities.

在空间中，线的单一、数组、延伸组构成面，辅以家饰的圆，线性中融入了非线性，方中有圆，圆中有方，亦是生活中理性与感性的结合。经由构图和色彩单纯的形式，体现秩序现象与均衡之美，艺术增加了空间的质感，并导入智能式生活，打造一个智慧居家及自然舒适的环境。

In space, the linearity, matrix and extension of lines make surface, and when complemented by furniture, linearity is fused with non-linearity, as whether it be a circle contained within a square or vice versa, it is a union of rationality and sensibility in life.Through simple elements such as composition and colors, it articulates the beauty of ordered phenomenon and balance, as arts elevate the quality of space with smart living adopted, it creates a smart residential environment with natural comfort.

东方之名，以合之意
The Name of the East, The Meaning of the Combination

项目地点：中国安徽/ Location：Anhui, China
项目面积：418 平方米 /Area：418 m^2
设 计 师：邹洪博 / Designer：Born Zou

IAI 设计优胜奖

邹洪博 Born Zou

曾为多个高端住宅、酒店项目担纲主创设计，拥有十多年的室内设计经验。作品提倡人与空间的互动，追求创新，强调品味、舒适的空间环境。他希望可以通过多年积累的设计经验，为国内的客户提供国际化的设计理念及优质的设计服务，给室内设计行业带入一股新的能量。

Born Zou has worked as a master designer for several high-end residential and hotel projects and has more than 10 years of interior design experience. The work promotes the interaction between people and space, pursues innovation, and emphasizes the taste and comfort of the space environment. He hopes that through years of accumulated design experience, he will provide domestic customers with international design concepts and high-quality design services, bringing a new energy to the interior design industry.

设计师用现代手法诠释东方之美，用新中式语言描述当代生活格调，用精致的材料与艺术品表现优雅、高贵的空间韵脚，演绎东方设计之美。

玄关连接着客厅和餐厅，通过垭口划分区域空间。精致的玄关摆件，与皮雕造型墙的层叠相得益彰。浮动繁花、山峦造势与独特的边柜，演绎出中式的情调。

The designer interprets the beauty of the East in a modern way, describes the contemporary life style in a new Chinese language, expresses the elegance and noble space rhythm with exquisite materials and works of art, and interprets the beauty of oriental design.
The entrance is connected to the living room and dining room, and the space is divided by the cornice. Exquisite porch ornaments brings out the best in each other with the cascading of leather-carved walls. Floating flowers, momentum built up by mountains and unique side cabinets create a Chinese-style atmosphere.

Public Buildings

公共建筑

张家界玻璃桥
Zhangjiajie Glass Bridge Hunan

项目地点：中国湖南 / Location : Hunan, China
设 计 师：渡唐海 / Designer : Haim Dotan

IAI 评审团特别大奖

渡唐海 Haim Dotan

国际建筑师、教育家和诗人，毕业于美国洛杉矶南加里福尼亚大学，获得了学士和硕士学位。
渡堂海教授是一位先锋建筑师，研究发展住宅、商业、工业、教育和公共机构的建筑施工方法。他在自己的项目中创造了一种新的国际建筑领域语言。

International architect, educator and poet, graduated from the University of Southern California, Los Angeles with a bachelor's and master's degree.
Professor Haim Dotan is a pioneering architect who studies the development of residential, commercial, industrial, educational and public institutions. He created a new language in the field of international architecture in his own projects.

“作为设计师，置身于仙境般的国家公园，我相信自然、和谐、平衡与美丽。自然本身就是美丽的，我们要尽可能地不去打扰她。因此，张家界玻璃桥采用了尽可能隐形的设计，使它自然而然地消失在白云中。”

正如老子所言：“大音希声，大象无形。”

这是一座水平的大桥，两边通过悬索固定在山上，桥身完全通透，并有扶手保护。人们站在桥上感受自然的纯粹，仿佛悬浮于空中，置身于天地之间，又如鸟儿展翅翱翔。

桥身距谷底几百米，是一座名副其实的奇幻之桥，勇气之桥。我更愿意叫它“勇敢的心之桥”，一首小诗聊以抒情：“飘零于空中，敞开心扉，我爱你。”

“As the designer of this bridge, located in an incredible and magical national park,I believe in nature, harmony, balance and beauty. Nature is beautiful as is. One wants to make the least impact upon it. Therefore, the Zhangjiajie Glass Bridge was designed to be invisible as possible--a white bridge disappearing into the clouds”.
As the ancient Chinese Dao Master, Lao Tzu, says: “Great sound is unheard;Great form is invisible.”
A horizontal bridge incorporatinga transparent glass floor with handrails and side suspension cables, it will create an experience of being in pure nature while suspended in mid-air, between heaven and earth, like a bird with its wings open wide.
Hundreds of meters above a canyon, it is definitely a Bridge of Wonder, a Bridge of Courage. I called it "Bridge of Courageous Hearts" and wrote this poem: “Floating in mid-air Open hearts – I love you.”

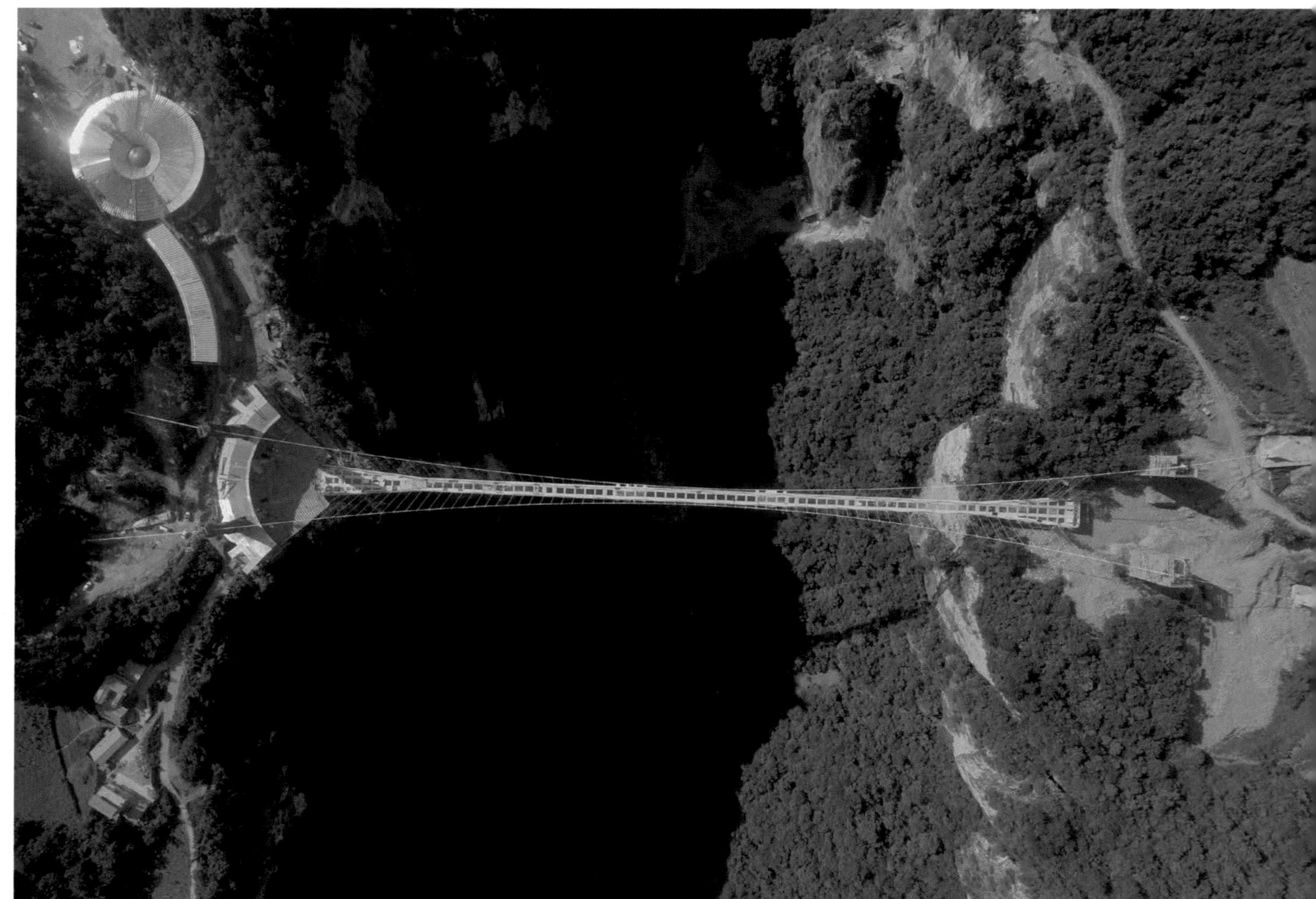

张家界玻璃桥是一座中等跨度的悬索桥，
所处喀斯特地貌地区，
桥两端的山谷中发现无数断层结构与地陷现象。

该桥两岸峡谷跨度为 385 米，
两边用于支撑悬索的桥塔相距 430 米。

为了融入自然以及实现主梁轻薄化的目标，主梁的梁高设计为 0.6 米，
因此该桥的高跨比达到了 625（375/0.6），
打破了中国的纪录。

收到了诗意建筑的启发，
中国首席桥梁工程师这样形容这座桥：

薄如翼，轻如燕，浮于空中。

主梁纵向两侧有钢箱结构，中间搭建横梁，与纵向钢箱结构交错相连。
这一结构与钢筋水泥板相似，但中间替代为玻璃，两边采不锈钢扶手。
两侧悬索展开像蝴蝶。

大桥工程基本数据

- 桥高 **300** 米
- 跳板台高 **280** 米
- 跳板台长 **385** 米（两座塔桥之间相距 430 米）
- 桥宽 **6.0** 米
- 最大容载：**800** 人同时
- 最大承重 **2200** 吨（悬索除外）
- 项目持续时间：**2011 - 2016**

GENERAL DATA

- Bridge height **300** meters
- Bridge jumping height **280** meters
- Bridge length **385** meters
- Bridge width: **6.0** meters
- Maximum Occupancy: **800** people
- Physical Weight **2200** tons
- Project Duration: **2011 - 2016**

The *Zhangjiajie* Glass Bridge is a medium-span suspension bridge. The bridge is located in a typical Karst area. Many faults and sinkholes are found surrounding the bridge's two ends.

The canyon spans 385 meters and the tower which support the main cables stand on each side at a distance of 430 meters.

To achieve the architectural goal of integrating the bridge into nature and minimizing the height of the beams, a girder structure was designed with height of 0.6 meter and the span-height ratio reaches 625 (375 / 0.6), breaking the Chinese record.

Inspired by the poetic architectural design of the bridge, the chief Chinese bridge engineer described the girder in these graceful words, "as thin as a wing, and as light as a swallow, resembling a light slice floating in the sky".

刚艺红木家具体验馆
Gangyi Rosewood Furniture Experience Pavilion

项目地点：中国广东/ Location : Guangdong, China
项目面积：7400 平方米 / Area : 7400 m²
设 计 师：黄永才、蔡立东 / Designer : Roy Wong , Cai Lidong

IAI 评审团特别大奖

项目由独立的单体建筑组成，在保留城市肌理过的刚性几何形态下，通过建筑形态上意想不到的线面折向来构成不规则的几何体。通过不同的界面切割，转折之后的视角将被无限延伸。

不规则的三角形几何体由轧花不锈钢拼接而成。材料表面的处理，通过天气环境的变化建筑，表皮也随之变化。材料的特性在不同的角度呈现不同的反射，随人的走动，建筑的形体及反射都在产生变化。

The project is composed of independent monolithic buildings. Under the preserving rigid geometry of the city's texture, irregular geometry is formed by the unexpected line-folding of the architectural form. Cutting through different interfaces, heaven and earth seem to become part of the building itself, and the perspective after the transition will be infinitely extended. The irregular triangular geometry is made of spliced stainless steel and the properties of the material reflect different reflections at different angles.

黄永才、蔡立东
Ray Wong
Cai Lidong

黄永才，共和都市创始人、创意总监，致力于以创意，富有想象的创造力设计驱动更多的品牌价值，创新大胆的设计为空间注入新的活力，期间作品获得多项顶级全球赛事奖项。

蔡立东，2002 年毕业于广州美术学院，荣获 2006 年中国室内设计双年展优秀奖、2006 年第二届广东室内设计大奖铜、2012 年中国精英设计师、中国建筑及室内高级设计师等多项荣誉。

Roy Wong, founder & creative director of Gong He City, is committed to driving more brand value with creative and imaginative creativity. The innovative and bold design injects new vitality into the space, and the works won several top global competition awards.many honors such as Chinese architects and interior senior designers.

Cai Lidong graduated from the Guangzhou Academy of Fine Arts in 2002 and won the 2006 China Interior Design Biennale Excellence Award, the 2006 Second Guangdong Interior Design Award Copper, and the 2012 China Elite Designer.

鲜明而具有导向的大阶梯隐藏于建筑体块缝隙间的入口之中，却又显而易见。在通透玻璃的立面处设置水池的处理，水元素的引用为建筑的构成增添了自然的乐趣，继而把室内与室外的鲜活结合了起来。微风吹动，水面泛起涟漪，在阳光的照射下，跃动的水纹倒映在天花板上，建筑的倒影在水面，自然的新生气息流动在室内的各个角落，给人以平和宁静之感，活跃气氛，增添空间情趣。

在室内的空间上以建筑倾斜元素为拓展，将内部空间串联起来，方便人们在不同的角度体验空间和方便大体量的展品运输，顺利打通设计与生活的紧致性，同时满足对美学的想象。

The vivid and guided large ladder is hidden in the entrance between the crevices of the building block,but it's obvious.The pool is set below the facade of the transparent glass,the reference to the water element adds natural pleasure to the composition of the building,and then combines the liveliness of indoor and outdoor.The breeze blows and the water surface ripples,the vibrant water wave reflected in the ceiling,the reflection of the building is on the surface of the water when they are under the sunlight, the natural fresh air flows in every corner of the room, giving people a feeling of peace and tranquility,which makes the space vitally and interesting.

Extend the indoor space with architecturally inclined elements and connect the interior spaces in series to facilitate people to experience space at different angles and facilitate the transportation of exhibits in large quantities. The smoothness of design and life can be smoothly achieved while satisfying the imagination of aesthetics.

顶部的两个大的中庭把室外的光线引入室内，光影和时间较量着，瞬息万变。“天井”是通过传统的认知方式来“景观内移”，空间上也得以延展，设计师留出来的天井空间，在采光、通风等方面都得到改善；运用通透的材质，展现了室外对室内自上而下的积极美好，让我们对周遭的一切事物变得更敏感，善待自然赋予的能量，以更宏观的角度看待建筑、室内与人的根本关系，用简单的手法演绎具有多样性的有趣空间，动态的生命将让沉寂的建筑体运动起来。

The two large atriums are at the top ,which let outdoor light into the room, lights and time are contested and change rapidly.the courtyard is a display of “landscape shifting” through traditional cognitive methods,this allows space to be extended, and the space of the courtyard reserved by designers has been improved in terms of lighting and ventilation. Designers use transparent materials to shows the positiveness of outdoor to indoor top-down, making us more sensitive to everything around us,we need to be kind to the energy given by nature, and we must look at the fundamental relationship between architecture, interior and people from a more macro perspective, and designers using simple techniques to interpret the interesting space with diversity, dynamic life will make the silent building move.

华翔堂社区中心
Huaxiang Tang Community Centre

项目地点：瑞士 / Location : Switzerland
项目面积：4500 平方米 /Area : 4500 m²
公司名称：德科乌维·蒙许工作室 / Organization Name : Dirk U. Moench

IAI 评审团特别大奖

INUCE
ARCHITECTURE DESIGN URBANISM • Dirk U. Moench

德科乌维·蒙许是德国裔巴西建筑师，是建筑工作室INUCE的创始人。德科乌维·蒙许在中国和瑞士设有办事处。 德科的工作特点是他对当前的城市化速度对城市居民的影响表示关注，特别是在快速增长的发展中国家。 他在书中表示希望，建筑可以通过加强人与人之间的联系，发展当地遗产并促进精神体验，从而在受灾的城市环境中引发局部修复过程，从而满足人类最基本的需求。 在其完成之前，他的项目已获得国际赞誉，并已被广泛出版。 2019年5月，欧洲建筑中心与芝加哥庙会合作，为德科乌维·蒙许颁发了“欧洲40岁及以下”设计奖，使他跻身40岁以下欧洲40位最有前途的新兴建筑师之列。

Dirk U. Moench is INUCE•Dirk U. Moench, the founder of the Brazilian architect and architectural studio, with offices in China and Switzerland. Dirk' s work is characterized by his focus on the impact of current urbanization on urban dwellers, especially in fast-growing developing countries. In May 2019, the European Architecture Center teamed up with Chicago Athenaeum to present Dirk U. Moench with the "Europe 40under40" award, which listed him as one of the 40 most promising emerging architects in Europe under 40.

过去的记忆，对未来的希望。当福州的华翔教堂建于1938年时，它的尖塔是唯一一个从传统明式住宅的海洋中出现的垂直结构。然而，在过去的30年里，福州市中心经历了一个戏剧性的城市化进程，风景如画的天际线及其独特的屋顶几乎完全消失了。今天，小教堂发现自己处于一个真正的像迷宫一样的商场和办公大楼的底层。由于迫切需要额外的空间，华翔会众决定在历史建筑附近的土地上建立一个支持性的社区中心。

A Memory from the Past, a Hope for the Future:When Fuzhou' s Huaxiang Church was built in 1938, its steeple was the only vertical structure emerging from an ocean of traditional Ming-style residences. During the last three decades, however, the city centre of Fuzhou has undergone a dramatic process of urbanization, and the picturesque skyline with its characteristic roofs has almost entirely vanished. Today, the little church finds itself at the bottom of a veritable maze of shopping malls and office blocks. In dire need for additional spaces the congregation of Huaxiang decided to build a supporting community centre on a plot of land adjacent to the historical building.

在风暴之眼：对该项目施加的挑战非同寻常。首先，功能和空间要求与遗产管理局对历史保护区施加的高度和面积限制相冲突。其次，四面环绕着障碍物，公共街道几乎看不到建筑物。第三，旧教堂的邻接和高低、现代和传统、东方和西方建筑的低交流环境，要求对新旧建筑的关系采取明确的态度。

In the Eye of The Storm:The challenges imposed on the project were extraordinary. Firstly, the functional and spatial requirements were conflicting with height and area restrictions imposed by the heritage authority onto the historically protected area. Secondly, surrounded by obstacles on all sides, the building would hardly be visible from the public streets. Thirdly, the adjacency to the old church and an uncommunicative environment of buildings high and low, modern and traditional, Eastern and Western, demanded a clear attitude regarding the relationship of old and new architecture.

武汉东原售楼处
Light Waterfall

项目地点：中国武汉 / Location : Wuhan，China
项目面积：1700 平方米 /Area :1700 m^2
设 计 师：凌子达 / Designer : Kris Lin

IAI 评审团特别大奖

凌子达 Kris Lin

取得法国CNAM建筑管理硕士学位。2001年在上海成立了“KLID达观国际设计事务所”。2006 出版个人作品集《达观视界》。共累计荣获全球奖项692项;在A'设计奖获奖设计师排行榜、世界设计排行榜中，凭借210分的好成绩,获得全球第三名，华人中国第一名殊荣由International Design Awards的Winner Ranking统计，凭借1400分的总分，在全球获奖设计师中获得“全球第八名”和室内建筑部分“中国第一名”的殊荣。

Received a master's degree in building management from CNAM, France. In 2001, "KLID Daguan International Design Office" was established in Shanghai.
2006 Published a collection of personal works "Da Guan Vision".
A total of 692 global awards have been won; In the A'design award award-winning designer rankings and world design rankings, with a score of 210 points, the third place in the world, the Chinese first place in China by the International Design Awards Winner Ranking statistics, with a total of 1400 points Points, won the [World's eighth place], indoor building part [China's first place] in the world's award-winning designers

这是一个以水为主题的艺术形式展示空间，设计师企图把室内、建筑、景观整合成一体，运用不同的方式来呈现水的状态。比如光的方式来呈现水瀑布流体的动态，和景观水池内真实流动的水相呼应和结合，让人们感受到以不同的水所呈现的方式来设计一个艺术形式的展示空间。

水是本案的主题，建筑设计上以通透的意念来传达了水和建筑、水和自然的关系。室内为展示区和洽谈区，运用了具有几何感立体格栅构成了墙面进行空间的隔断。洽谈区旁边是由连续的玻璃墙面环绕水池而成的中庭景观，成为了室内的一个景观点。

水是本案的主题，设计上以通透的意念来传达了水和建筑、水和自然的关系。企图利用光的形式做出水瀑布的效果并形成一个瀑布的流体感；运用LED灯、木作材料等设计精致的节点，进一步呈现光的水幕墙造型。室内空间通过玻璃墙面环绕水池而成的中庭景观，成为了室内的一个景观点。

This is an art form showing space with water as the theme. The designer tries to integrate the interior, architecture and landscape into a whole and present the state of water in different ways. Presented such as light fluid dynamic water falls, and the real flow of water in the landscape pool as a photograph echo and combine, let people feel in different water present ways to design an art form of display space.

Water is the subject of this case. The architectural design conveys the relationship between water and architecture, water and nature with the idea of transparency. The interior is the exhibition area and the negotiation area, and the partition of the space on the wall is formed by using the three-dimensional grille with geometric sense. Next to the negotiation area is the atrium landscape formed by the continuous glass wall surrounding the pool, which becomes an indoor landscape point.

Water is the subject of this case, which is designed to convey the relationship between water and architecture, water and nature with the idea of transparency. We attempt to use the form of light to create the effect of a water waterfall and form a sense of the flow of a waterfall; LED lights, wood materials and other exquisite nodes are used to further present the water curtain wall shape of light. The atrium landscape formed by the glass wall surrounding the pool becomes an indoor landscape point.

街
Street

项目地点：中国武汉 / Location : Wuhan, China
项目面积：200平方米/ Area : 200 m²
设 计 师：汤 晶 / Designer : Tang Jing

IAI 设计优胜奖

汤晶 Tang Jing

湖北美术学院硕士毕业中级工程师
现任湖北广播电视大学环境艺术设计教研室讲师
寰屹零贰设计工作室创始人
2018 年 IAI 设计大赛个人组优胜奖
武汉市青年设计师协会资深会员

Master of Engineering, Hubei Academy of Fine Arts
Hubei Broadcasting and Television University
Department of Environmental Art Design, Lecturer
Founder of the Huan Yi Ling Er Design Studio
2018 IAI Design Award Individual Group Excellence
Senior member of Wuhan Youth Designers Association

该项目为实际设计项目，改造成步行街，建筑风格属于殖民建筑，也称为民国建筑，具有西式建筑的外立面、中式建筑的内墙结构。属于武汉市武昌首义文化区的其中一块。设计要求主要是保留一部分完好的民国建筑，修复一部分，拆迁一部分危险性房屋。保存老建筑风貌，设计新式的建筑外形和使用环境，辞旧迎新，变的是形式，不变的是情怀。

The project is a practical design project, which is transformed into a walking street. The architectural style belongs to the colonial architecture, also known as the Republic of China, the external facade of the western style architecture, and the inner wall structure of the Chinese style architecture. It belongs to one of the first cultural districts in Wuchang, Wuhan. The main requirement of the design is to preserve some of the intact buildings in the Republic of China, repair part of them, and remove some dangerous houses. Preserving the style of the old buildings, designing the new architectural appearance and using environment, and changing the old into the new are changing form and unchanging feelings.

民主路西段步行街改造方案
The west of Minzu road landscape design

龙湖原麓社区中心
Longfor Yuanlu Community Cente

项目地点：中国重庆/ Location : Chongqing,China
项目面积：4000 平方米 /Area : 4000 m^2
公司名称：成执设计 / Organization Name : Challenge Design

IAI 设计优胜奖

成执设计 Challenge Design

上海执琢建筑设计事务所有限公司（建筑设计事务所甲级）公司自创办以来始终以"精细化设计，专业化服务"为宗旨，主持设计的作品类型包含了"商业综合体""度假社区""高端住宅""精品酒店""文化艺术"等项目 受到了业主及行业的一致好评。

The Aim Of Challenge Design Pte Ltd (Class A Architectural Design Office) Has Been To Provide "Refined Design And Professional Service" Since Its Foundation. The Company Has Undertaken Projects Ranging From "Commercial Complex", "Resort", "Luxury Condo", "Boutique Hotel", "Museums", And "Public Architectures, Winning Dozens Of Awards And Unanimous High Appraisal.

项目毗邻龙兴古镇，御临河边，有远山起伏，绿化集中，东侧景观极佳，而其余方向则乏善可陈。

作为对环境的回应，建筑师将三个体量不同的盒子并列于山坡上，垂直于滨河道路，以期达到滨河景观及远山的最佳观赏。通过建筑与环境的互塑共生，使远山、河流、庭院、建筑重塑构成新的秩序，营造独有坡地特征的建筑群落。

Facing Yulin River, the Project sits next to Longxing Ancient Town, Chongqing, China, mirrored by rolling hills in the distance and a centralized green area. Natural landscape on the east side is exceedingly fascinating while its counterparts on the other sides are relatively ordinary.
As a response to the environment, the architect places three buildings of different sizes side-by-side on the hillside and vertical to the riverside road, looking to achieve the optimum visual effect of riverfront landscape and distant mountains. The best view of the distant mountains. Through the mutual shaping and symbiosis between architecture and environment, the mountains, rivers, courtyards and buildings will be reshaped to form a new order, and the architectural community with unique slope characteristics will be created.

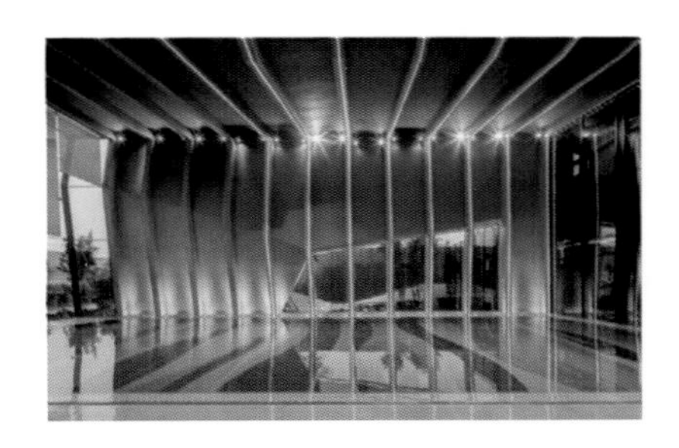

Architectural Conceptual Design

建筑概念设计

三叠泉
San Die quan

项目地点：中国江西 / Location : Jiangxi, China
项目面积：4778 平方米 /Area : 4778 m^2
公司名称：创空间集团 / Organization Name : Creative Group

IAI最佳设计大奖

創空間
CREATIVE GROUP

创空间 Creative Group

汇集建筑与室内设计背景的多元人才，配合组织化管理模式，将工程进度做系统化的全面控管，
提供从前期沟通、设计、合约拟定到后期执行的完整服务。2000 年成立“权释设计”。2014 年，2015 年陆续成立“CONCEPT北欧建筑”与“JA建筑旅人”建筑设计公司，传递人与自然和谐共生的理念精神，秉持着“提升国人生活品质、共创生活美学体验”为目标持续优化。

Bringing together diverse talents in the background of architecture and interior design, and coordinating the organizational management model, systematically and comprehensively control the progress of the project, providing complete services from pre-communication, design, contract formulation to post-execution. In 2000, the company established the "ALLNESS Design" . In 2015, it established the "CONCEPT " and the "Journey Architecture" architectural design company to convey the spirit of harmony between Humanity and Nature Environment, and uphold the "quality of life of the people." is the continuously optimized goal .

三叠泉，来自江西的深山，以层层叠加的瀑布著名。中国著名诗人苏东坡曾在此一游，并诗句：“不识庐山真面目，只缘身在此山中。”中国人深信，天无绝人之处，只要改以不同角考人生，便能豁然开朗。因此借助庐山的形象，将诗的意象量体化，在建筑中表达出东方人的视角而豁达的人生哲理。隐：建筑体“隐”藏在山体中，使人自世俗中转换心境，如同古人居，脱离世俗的纷扰，能真正在山中享受宁静”“转换视角”，便能看到一片新的天地。叠：体的层层叠加，与垂直错位，转换空间思考的角度，引导人转换思维，能带给居住者全然不同境。泉：透过格栅打造出泉水流泄而下的意象，透过光与影在不同的时分，打造出流动性，同具有保护隐私的功能。

菜天下
H-Vall

项目地点：中国哈尔滨 / Location : Harbin, China
项目面积：416100 平方米 /Area : 416100 m^2
公司名称：北京艾奕空间设计有限公司 / Organization Name : AE Archintects

IAI最佳设计机构大奖

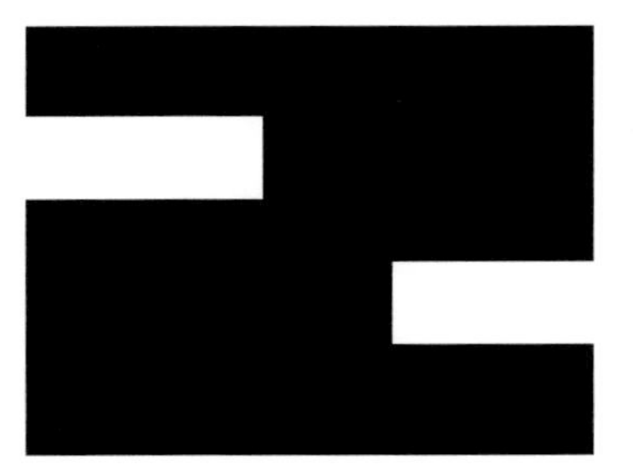

AE ARCHITECTS 艾奕设计

艾奕设计 AE Architects

艾奕空间设计创立于北京，得益于创始人Abel Erazo 在建筑设计、景观设计以及室内设计领域二十多年的实践经历。我们的艺术哲学观受古代东方禅宗和西方现代化摩登的影响，形成了简约、朴素、纯洁、象征主义的风格,自成一派。我们拥有丰富的餐饮、办公、酒店、住房、商业和文化设计经验，确保创造高效、舒适和独特的空间。

AE Architects was founded in Beijing and establish its practice from the 20 years of experience of Abel Erazo in Architecture, Landscape and Interior Design. Our philosophy has been influenced by Ancient Asian Zen and Western Modern. Taking from both the ideas of simplicity, pureness and symbolism. With vast experience on food and beverage, office, hospitality, housing, commercial and cultural design, we ensure the creation of efficient, comfortable and unique spaces.

哈尔滨菜天下项目的设计灵感是来自于这个独特的地方。哈尔滨位于中国东北边境，每年的冬天以冰雪节和白色雪景著称。而冰晶中的美和寒冷的天气，正是市民们所引以为傲的地方。当雪花被放大后，我们发现了每朵雪花的冰晶结构都是不尽相同的。每片单晶都是独一无二的，这使它们变得非常神秘，流露着含蓄的美。我们的目标是让它被更多的人所看到，并为整个城市创造一个地标，可以吸引来自全省、整个国家甚至是世界各地的游客。

The design inspiration for the Harbin V-Hall project comes from this unique place. Located on the northeastern border of China, Harbin is known for its snow and ice and white snow every winter. The beauty and cold weather in the ice crystals are the places that the citizens are proud of. When the snowflake was enlarged, we found that the ice crystal structure of each snowflake is not the same. Each single crystal is unique, which makes them very unique and mysterious, revealing subtle beauty. Our goal is to make it visible to more people and to create a landmark for the entire city that attracts visitors from across the province, the entire country and even the world.

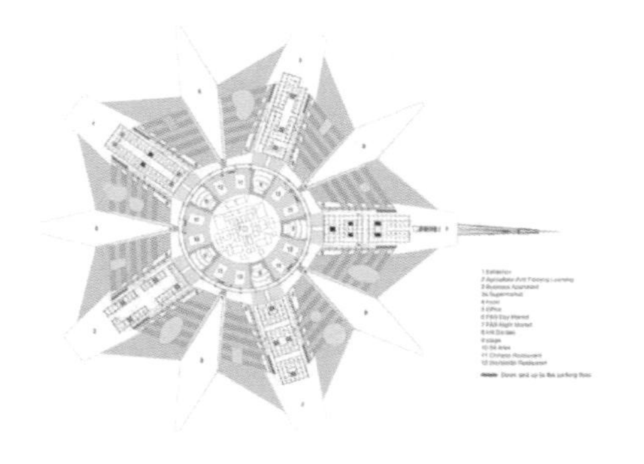

Renovation & Reconstruction of Old Buildings

旧建筑改造与重建

倾城客栈
King Charm

项目地点：中国台湾 / Location：Taiwan, China
项目面积：235 平方米 /Area：235 m²
设 计 师：李硕 / Designer：Li Shuo

李硕 Li Shuo

北京水石空间环境设计有限公司联合创始人、S+STTJ 设计工作室创意设计师。
经过 18 年的时间历练，也随着经验和眼界的增长与积累，从艺术与社会的角度对设计有了更深层次的、更根本的理解。

Beijing's Shui and Shi Space Environment Design Co. Ltd. co-founder
S+STTJ design studio creative designer
After 18 years of experience, with the growth and accumulation of experience and vision,From the perspective of art and society, it has a deeper and more fundamental understanding.

在美丽、神秘的云南腾冲，有一片水草丰美、气候宜人的湿地——北海湿地，而在北海湿地的东岸，有一个清新、雅致、供人冥想的空间——倾城客栈。腾冲的建筑风格可追溯到清代，深受徽派建筑的影响，粉墙黛瓦，瘦柱肥梁，又加以腾冲天然特有的建筑材料——火山石，形成了独特的建筑风格。倾城客栈则是将传统的建筑元素提炼，将火山石用以新的拼装和组合方式呈现，并和钢筋、玻璃、实木相配合，形成了脱俗的禅与意的空间。在面向湿地及夕阳的方向设置了一个冥想空间，可在大型的圆环造型中央禅坐，纳天地之灵，放空自身。

In the beautiful and mysterious Tengchong of Yunnan, there is a beautiful and pleasant Wetland - the Beihai wetland, and on the east coast of the wetland of Beihai, there is a fresh, elegant and meditative space - the city inn. The architectural style of Tengchong can be traced back to the Qing Dynasty. It was deeply influenced by the architecture of Hui school. The whit wall and black tiles, thin column fertilizer beam, and the natural building material in Tengchong, the volcanic rocks, formed a unique architectural style. The inns are to refine the traditional architectural elements, to use the volcanic rocks for new assembly and combination, and to cooperate with steel, glass and solid wood to form the free space of Zen and meaning. In the direction of the wetland and sunset set set set a meditation space, can be in the center of a large circular shape meditation, the spirit of heaven and earth, empty itself.

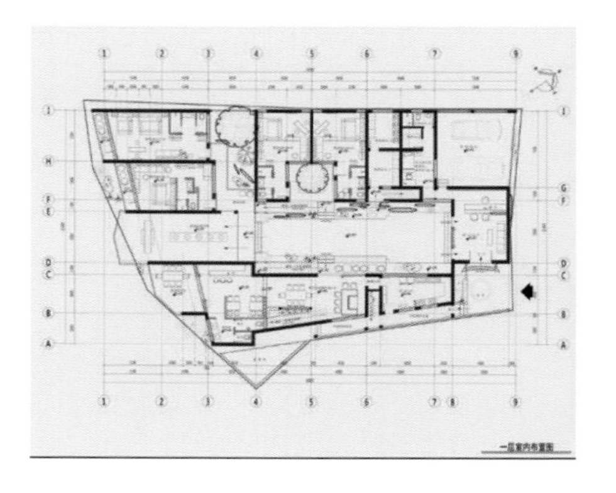

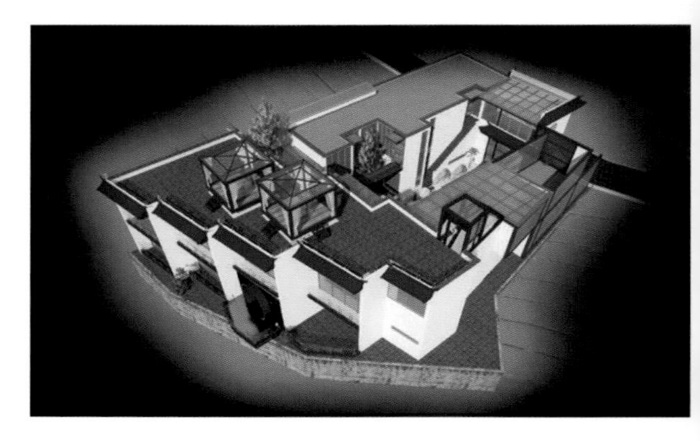

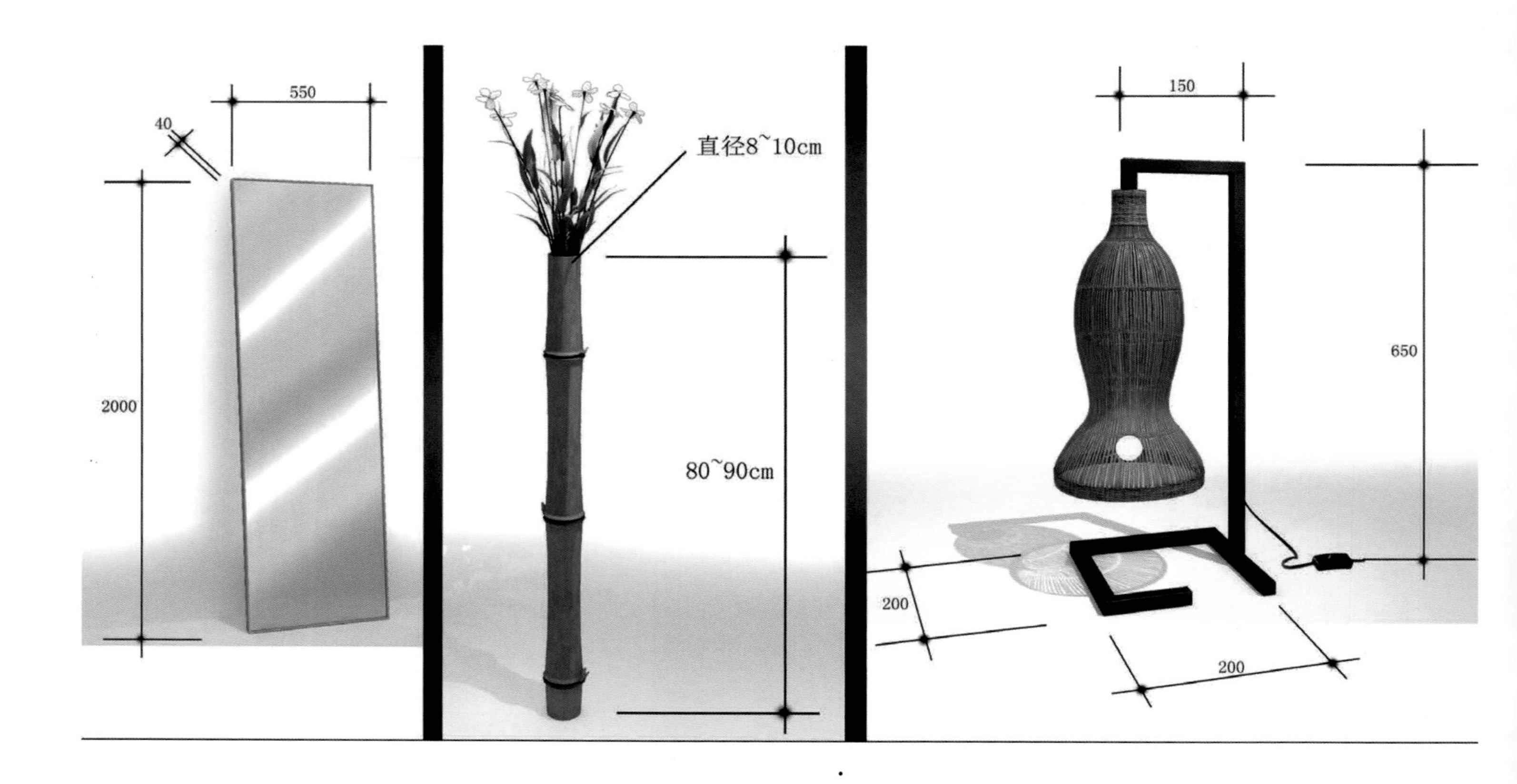

二楼顶还设有两个错落悬空的玻璃亭台，形成了“翠玲珑”的效果，可欣赏美景。客栈中设置前室、车库、工作用房、餐厅、茶室、客房、冥想空间、瑜伽平台及主人的私人空间，通过前室进入庭院后在通过各个通道与楼梯可到达，这样既区分了空间，又保证了各个空间的单独性与联系。本案是一个二层“回”字形的建筑形态，四周为建筑实体，中心为庭院，这个也尊重云南传统民居“三坊一照壁”形式，使之在自然的环境当中融入当地的建筑群落。而从总平面中，拱形的大门洞、飞檐及“回”字形的建筑，再到门口玄关的圆形地拼，整体形成一幅“龙吐珠”的形态，这在当地也是非常吉祥的寓意。

Two on the top of the building, there are two glass pavilions dangled and suspended, forming the effect of "Cui Linglong", which can appreciate beautiful scenery. The inn, the garage, the working room, the restaurant, the teahouse, the 9 rooms, the meditation space, the yoga platform, and the owner's private space can be reached through the various passages and staircases through the front room, which not only distinguishes space, but also ensures the separate new and connected space. This case is a two - story Chinese character " 回 " shape of the architectural form, around the building entity, the center of the courtyard, which also respect the traditional Yunnan folk house "Three - one - wall" form, so that the natural environment integrate the local building community. From the general plane, the arch gate hole, eaves, and the Chinese character " 回 " shape of the building, and then to the doorway of the round spelling, the whole form a "dragon beads" form, which is a very auspicious meaning in the local.

小溪家
Springstream House

项目地点：中国福建 / Location : Fujian, China
项目面积：275 平方米 / Area : 275 m²
公司名称：WEI 建筑设计事务所 / Organization Name : WEI Architects Elevation Workshop

IAI 最佳设计大奖

WEI 建筑设计事务所

WEI Architects Elevtion Workshop
最初在纽约成立，后于 2009 年正式成立北京事务所。作为在中国最活跃的年轻事务所之一，WEI 的作品已被包括《domus》、中国的《时代建筑》、美国的《Archdaily》、英国的《DeZeen》、荷兰的《FRAME》，德国的《Gestalten》、日本的《六耀社》等世界十几个国家的五十多个杂志和出版社报道，并获得国内外多个奖项。

Originally founded in New York, WEI architects (aka. ELEVATION WORKSHOP) established its Beijing office in 2009. As one of the most innovative and leading architecture firms in China, WEI architects has worked on projects of various scales. Positioned at the crossroads of art and architecture, WEI' s projects were featured in over fifty publications in over ten countries around the globe, including Time Architecture (China), Domus, Archdaily (U.S.), Dezeen (UK), FRAME (the Netherlands), Gestalten (Germany), and Rikuyosha (Japan). Several WEI' s projects have been nominated for international design awards.

房屋结构利用回收的老木料和部分新材料，由当地的师傅和村民，结合当地的建造方式建造。室内布局尊重当地以炉灶为核心的传统，保留很多当地做法的同时结合智能家居设备。厨房和茶室之间，采用了当地特殊的窗户设计，既可以关上成为墙板，又可以打开成为操作台面。同时设计了“会呼吸的窗”，利用当地季节性风向变化，加强对流，从而形成节能环保的被动式降温除湿的作用。房屋周边的景观利用当地植被，保持建筑与已有环境的衔接。

目前小溪家已经成为当地的标志性建筑。每个来到这里的人都能在建筑空间里找到属于自己的感动。

The structure of the house uses recycled old wood and some new materials, built by local masters and villagers, combined with local construction methods. The interior layout respects the local tradition of taking the stove as the core, retaining many local practices while incorporating smart home equipment. Between the kitchen and the tea room, the special window design of the local area is used, which can be closed as a wall panel and opened as a countertop. At the same time, the “window that will breathe” is designed to take advantage of local seasonal wind direction changes to enhance convection, thus forming a passive cooling and dehumidification effect of energy conservation and environmental protection. The landscape around the house uses local vegetation to keep the building connected to the existing environment.
At present, the Springstream House has become a local landmark, and everyone who comes here can find their own touch in the architectural space.

Product Design

产品设计

余乐园
My Playground

项目地点：中国杭州 / Location : Hangzhou , China
公司名称：噪咖艺术有限公司 / Organization Name : Noise Kitchen Art Co.,Ltd

IAI最佳设计机构大奖

噪咖艺术
Noise Kitchen Art Co.,Ltd.

噪咖艺术有限公司为新媒体艺术类策划团队，致力于发展新媒体艺术，将作品结合艺术与科技，并将两个不同领域的概念融入人们的日常生活，创作上习以富视觉刺激的装置呈现与引人入胜的论述观点，平衡艺术与科技的两端，让观者得以在资讯、体验、参观的过程兼获美感与科技性的洗礼，带领民众共享数位艺术生活乐趣。

NoiseKitchen is a team focusing on new media art.
noiseKitchen sets its goal to the development of installation creation, creative merchandises and promoting new media art; the results will be presented and seen in digital art exhibitions, public art works, educational events and cultural/creative spaces.

位于大陆杭州余杭的余之城生活广场，《余乐园》是一个科技艺术游乐园，其中包含《光击跷跷板》《城市摇摆》《心跳房子》《风车树》与《好水时节》《家聚》以及《余杭一号》7件互动艺术装置。不仅体现科技艺术对“风”“光”“水”等传统文化元素的当代转译，更以互动科技的装置招唤每个消费者的童年游乐记忆。

For the new media art planning team, NoiseKitchen Art Co.,Ltd. dedicated to the development of new media art, combining works with art and technology, and integrating the concepts of two different fields into people's daily life, creating a visually stimulating device with a compelling discussion The idea is to balance the two ends of art and technology, so that the viewer can enjoy the beauty and technology baptism in the process of information, experience and visit, and lead the people to share the joy of digital art life.

字由诗
Character by Poetry

作品尺寸: 255×255×30毫米 / Work size：255×255×30mm
设 计 师：易成海 / Designer : Yi Chenghai

IAI 最佳设计大奖

易成海 Yi Chenghai

中国优秀青年设计师
厦门易壹文化创意设计公司艺术总监
厦门有壹个空间设计事务所设计总监
"文化部文化产业创业创意重点人才库" 成员
设计石材量产产品三十多项，获国家授权发明专利一项、国家实用新型专利十余项，获国内外设计奖二十余项。

Chinese Outstanding Young Designer
Xiamen YIYi Cultural Creativity Design Company Artistic Director
Xiamen You Yi GE space design firm Creative director
Member of "Ministry of Culture, Cultural Industry, Entrepreneurship, Creative Key Talents"
Designed more than thirty mass producted stone works, obtained ten National authorized patent for utility model, won Twenty more items domestic and overseas design awards.

《字由诗》借用了中国传统活字印刷的活字雕刻手法，将茶盘与传统活字结合。通过在平面上采用不同材质的互动转换，单个活字的搭配组合，用现代简约的线条和有层次的造型，在满足使用者泡茶排水需求同时，促进使用者产生对中国传统文化的互动。

活字印刷作为中国古代四大发明之一，对世界文明进程和人类文化发展产生重大影响，是人类史上一次伟大的技术革命，迄今已经跨越千年。此茶盘借助乌金石与铜字的组合，打造一款属于一个人或两人对饮的茶盘，每一个茶盘都有一个故事，通过活字组合的形式呈现。同时采用活字元素组成有关茶的诗词，可以用宣纸拓印，并展现出独特的艺术效果。

"Character by poetry" borrows the Chinese tradition movable type carving technique, combine the tea tray with the tradition movable type character. Through the interactive transformation of different materials on the plane, the matching and combination of single type, the modern and simple lines and the hierarchical modeling, meet the user's needs of tea infusing and drainage and promote the user's interaction with traditional Chinese culture.

As one of the four great inventions of ancient China, movable type printing has had a significant impact on the progress of world civilization and the development of human culture. With the combination of Wujin stone and copper word, this tea plate is made into a tea plate which belongs to one person or two people drinking together. Each tea plate has a story, which is presented in the form of movable type combination. At the same time, the poetry related to tea is composed of movable type elements, which can be printed on rice paper and show a unique artistic effect.

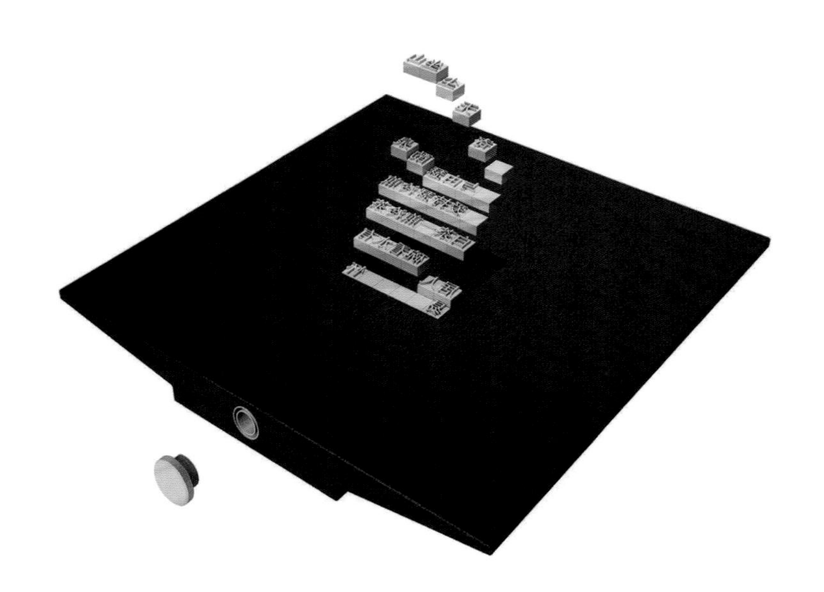

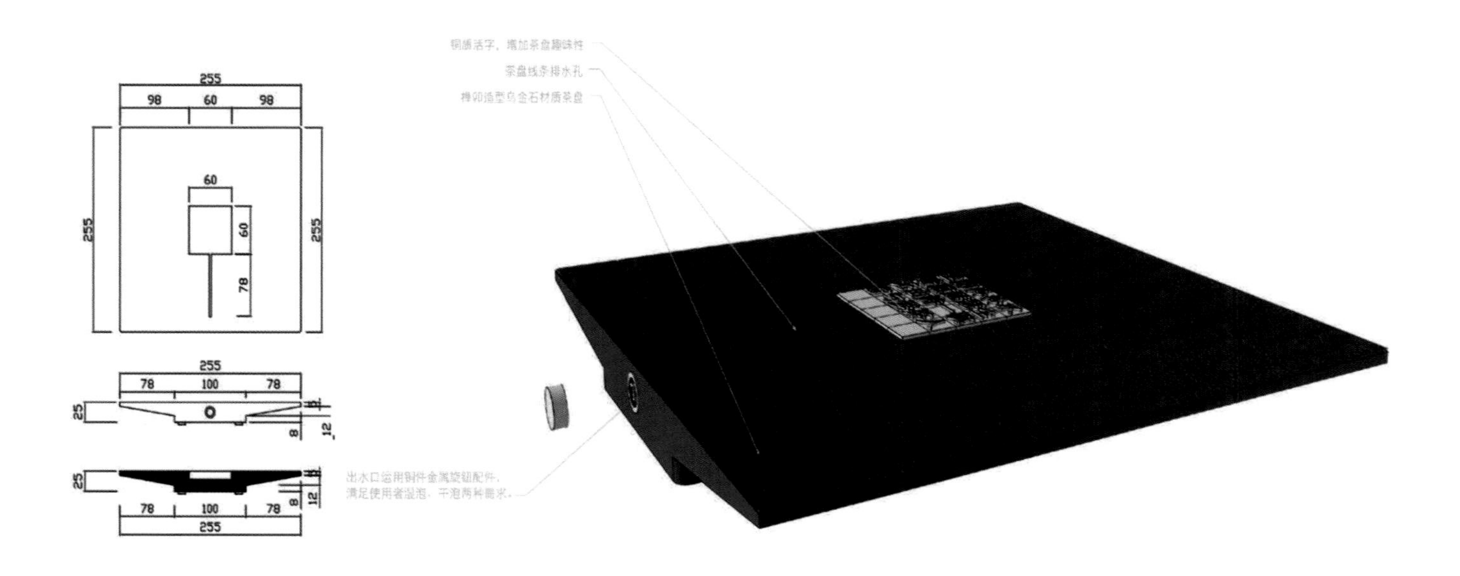
255
98
60
98
60
60
78
255
255
255
78
100
78
25
8
12
25
8
12
78
100
78
255
铜质活字，增加茶盘趣味性
茶盘线条排水孔
榫卯造型乌金石材质茶盘
出水口运用铜件金属旋钮配件，
满足使用者湿泡、干泡两种需求。

鳞灯系列
Scales lamps collection

设计师：戴维德·蒙塔纳罗 / Designer : Davide Montanaro

IAI最佳创新力设计大奖 **IAI评审团特别大奖**

戴维德 · 蒙塔纳罗
Davide Montanaro

戴维德 · 蒙塔纳罗，1972 年出生于库内奥，将他的整个一生奉献给工业设计。他的专业范围很广泛：从徒步鞋的连接系统设计到书柜的模块化系统，从灯具到展台，通过博物馆书店的产品。他的设计的独特之处在于研究和转移来自不同工业部门的材料和技术。2018 年 12 月，在上海著名的 IAI 设计奖中获得最佳产品、最佳创新力大奖。

Davide Montanaro, born in 1972 in Cuneo, has lived and worked in Milano since 2000, dedicating his whole activity to industrial design. His professional range is various: from the design of connection systems for trekking shoes to modular systems for bookcases, from lamps to exhibition stands, passing through products for museum bookshops. Distinctive features of his activity are the research and the transfer of materials and technologies from different industrial sectors. In December 2018 studiodsgn received the award for best product, best creativity and the critical award at the prestigious IAI Design Award in Shanghai.

鳞灯、钢灯系列：该灯的设计结合了光化学蚀刻的高生产技术和手工造型的不确定性。通过这种方式，从标准化的工业过程中，我们实现了形式的极端个性化，从系列转向特定。

该灯由钢板制成，厚度为2/10毫米，由化学切割的60X60厘米的板获得。每个灯具有不同的最终形状，因为它是手工模制的。墙上的阴影是一个非常有趣的组成部分。提供的卤素灯泡功率为60瓦。

该系列在装饰购物活动期间，在巴黎著名的卡鲁索广场展出。在米兰，它在XXI国际三年展期间展出。在约旦，这个系列在安曼的活动中展出。

Scale Light, Steel lamp collection:
the design of this lamp combines the high production technique of photochemical etching to the final indeterminacy of the shape, obtained manually.
In this way, from a standardized industrial process , we arrive at an extreme personalization of forms , passing from the series to the particular.
The lamp is made of steel sheet, thickness of 2 tenths of a millimeter, obtained from a plate of 60 to 60 cm chemically cut. Each lamp has a different final shape because it is molded by hand. Shadows on the wall are a very interesting component. The supplied halogen bulb has a power of 60 watts.
The collection was exhibited in the prestigious setting of the Carrousel du Louvre in Paris during the Deco Shopping event. In Milan it was displayed during the XXI International Triennale. In Jordan, the collection was presented in Amman at an event created by Hyper-room.

电子白噪音发生器
Electronic White Noise Sound Machine

公司名称：深圳前海帕拓逊网络技术有限公司 / Organization Name : Patozon

IAI 最佳产品设计大奖

Patozon

前海帕拓逊 Patozon

帕拓逊是国内 A 股上市企业跨境通的全资子公司，是全球消费者美好生活的物质提供商，以“点靓每个人的生活”为使命，为全球个人及家庭提供高品质的消费类产品，客户遍布全球三十多个国家和地区。帕拓逊也是国家认定的"国家高新技术企业”是集产品设计、研发、销售、服务为一体的高新科技公司。

Patuoxun is a wholly-owned subsidiary of the domestic A-share listed company Cross-border (A-share code: 002640). It is a material provider of a better life for consumers around the world, with the mission of “pointing everyone's life” as the global Individuals and families provide high-quality consumer products with customers in more than 30 countries and regions around the world. Patuoxun is also a national high-tech enterprise recognized by the state. It is a high-tech company integrating product design, research and development, sales and service.

它能确保给您个人一个更好的夜晚睡眠和安宁的休息。由AC或USB供电，在你选择的波浪、雨、瀑布或风暴声音中动态产生舒缓的白噪音。按钮在方便操作的位置可以自定义您选择的声音的音调和音量，以及自动关闭计时器。这款小巧便携的设计带有16度倾斜角，适合任何地方，不引人注目地融入周围环境，让您轻松自然地排除干扰噪音，在家里或旅途中享受一个健康的夜晚。它能提供一个理想的放松、学习、私语，或者任何您想要控制您声音的环境。所以您可以在任何地方生活或睡眠从而享受最美好的环境。

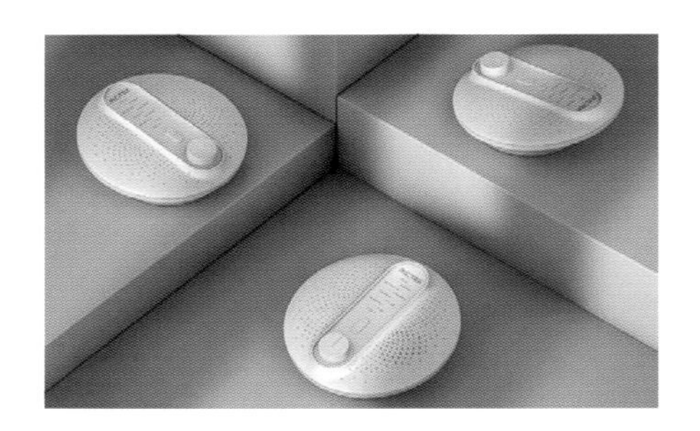

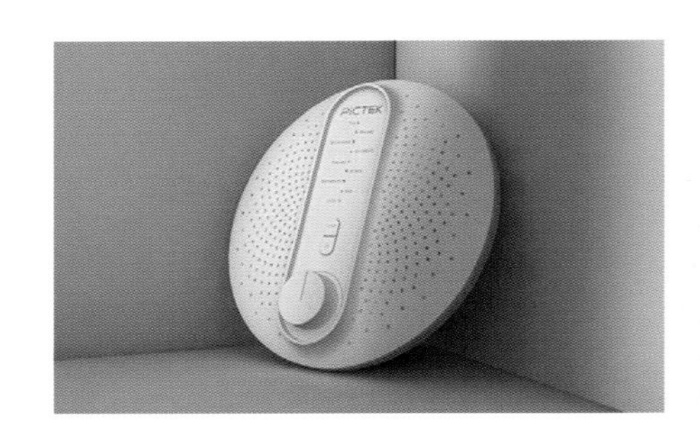

It’s designed to be your personal white noise sound machine to ensure a better night's sleep and peaceful rest. Powered by AC or USB and dynamically produces soothing white noise in your choice of waves, rain, waterfall, or storm. Buttons are obvious to customize the tone and volume of your chosen soundscape as well as the automatic shut-off timer. The compact and portable design with 16 degree tilt angle fits anywhere and blends unobtrusively into surroundings, makes it easy and natural to tune out distracting noises and enjoy a healthy night's sleep at home or on trips. It’s ideal for relaxation, study, speech privacy, or any situation where you'd like more control of your sound environment, so you can live (or sleep) your very best life uninterrupted-anywhere.

机械白噪音发生器
Mechanical White Noise Sound Machine

公司名称：深圳前海帕拓逊网络技术有限公司 /Organization Name : Patozon

IAI 最佳产品设计大奖

创新的白噪音发生器设计灵感来自风扇叶片，顶部和侧壁通风口十分具有特色。它可以模拟128种自然白色风噪声，旋转风扇叶片由1赫兹旋转、无刷直流电机驱动，产生自然的无声无噪音的声音保证了高质量及舒适的睡眠。顶部的可调速度旋钮还可以让您微调白噪声的音调。另外，LED灯增强了平和与宁静的氛围。在睡觉时，额外的USB充电端口可以同时为任何设备充电。它采用斯堪的纳维亚灰色色调的时尚设计，使其适合任何家居环境或随身携带。

The creative white noise generator design inspired by a fan with vanes and blades featuring on the top of surface and side wall vents. It can mimic 128 types of natural white noise of wind blow by a rotating fan blade spinning at 1Hz which powered by brushless DC motor, producing natural sound noise-less for good quality and comfort sleep. Speed adjustable knob at the top also allows you to fine-tune the tone of your white noise. Plus, a 5500K soft breathing LED light enhances better atmosphere of peace and quiet. An extra USB charging port can charge any device at the beside while you sleep. Its compact and stylish design with Scandinavian grey tone color makes it fit any furniture, or a carry-on bag.

智能宠物喂食器
Smart Pet Feeder

设 计 师：熊沐国 / Designer : Xiong Muguo

IAI 最佳设计大奖

这是一款智能宠物喂食器，内置无线模块连接网络，可以通过手机软件控制，如即时查看宠物及环境的视频、语音对讲以及投放食物等。外观造型特别考虑了防止宠物撞伤、防碰倒等功能性设计，同时解决了同类产品的卡粮问题，并能够通过手机较准确地控制投放食量。

This is an Smart Pet Feeder, which has a built-in wireless Wi-fi module connected to the network. It can be controlled by mobile app, such as instantly viewing pets and environmental videos, voice intercom and food putting in. And the function design , such as collision prevention and injury ,are especially considered in the appearance design.At the same time, it solves the problem of the card food of similar products, and can control the food intake through the mobile phone with more accurate control.

熊沐国 Xiong Muguo

迄今从业 15 年，发表学术论文若干，设计作品曾获得德国绿色产品设计奖、iF 设计奖、红点奖、美国 IDEA 产品设计奖、台湾金点设计奖、IAI 设计奖、红星奖、省长杯、醒狮杯等并带领团队获得红点奖、红星奖等国内外设计奖项。

He has been working for 15 years and has published several academic papers. His design works have won the German Green Product Design Award, iF Design Award, Red Dot Award, American IDEA Product Design Award, Taiwan Golden Point Design Award, IAI Design Award, Red Star Award, Governor's Cup. , lion cup and so on and led the team to receive red dot awards, red star awards and other domestic and international design awards.

木语——扁平化（榫卯）家具系列
Wooden language——Flat packaging (tenon mortise) furniture serie

设计师：翟伟民 / Designer : Zhai Weimin

IAI设计之星奖 IAI最佳设计大奖

翟伟民 Zhai Weimin

中国建筑学会会员
清华大学文创园区签约设计师
法国创意设计专业委员会会员
南京粽角文化创意设计有限公司创始人
台湾实践大学工业产品设计学系研究所在读

Member of China Architectural Society
Contractual Designer of Tsinghua University Wenchuang Park
Member of French Creative Design Professional Committee
Founder of Nanjing Zongjiao Cultural Creative Design Co., Ltd.
Research Institute of Industrial Product Design Faculty , Taiwan Practical University

此款衣帽架完全可以实现扁平化，整个造型以线塑造，结构上汲取了传统明式家具的榫卯结构，并合理演绎，运用在衣帽架上，顶部通过十字榫和夹头榫结构牢牢锁住上面的挂杆，下面采用十字榫和走马销结构，方便拆分，同时锁住下面腿部结构，上下达到结构力学的平衡。挂杆上翘，也便于挂放衣物或包之类物品。简洁实用，整个拆分过程也是传达了物与人的情感交流。扁平化椅子设计，便于网络电商平台销售，在生产与运输包装过程中扁平化衣可以节省大量成本，在用户体验方面，扁平化非常实用，同时便于携带和收纳。

The flat design allows for more interaction with people and space. This coat rack can be completely flattened, the whole shape is shaped by lines, the structure is drawn from the traditional Ming furniture, and it is reasonably interpreted. It is applied to the coat rack and the top is passed through the crossbow and the collet. Firmly lock the hanging rod above, the bottom is made of cross shovel and the horse-shoulder structure, which is convenient for splitting, and at the same time locks the structure of the lower leg, up to the balance of structural mechanics. The hanging rod is upturned, and it is also convenient to hang clothes or bags and the like. Simple and practical, the entire split process also conveys the emotional communication between things and people. The flat chair design facilitates the sale of the network e-commerce platform. The flat-coating can save a lot of cost in the production and transportation packaging process. In terms of user experience, the flattening is very practical, and it is easy to carry and store.

木 语

扁平化（榫卯）家具系列

Flat packaging (tenon mortise) furniture series

木 语

扁平化（榫卯）家具系列

Flat packaging (tenon mortise) furniture series

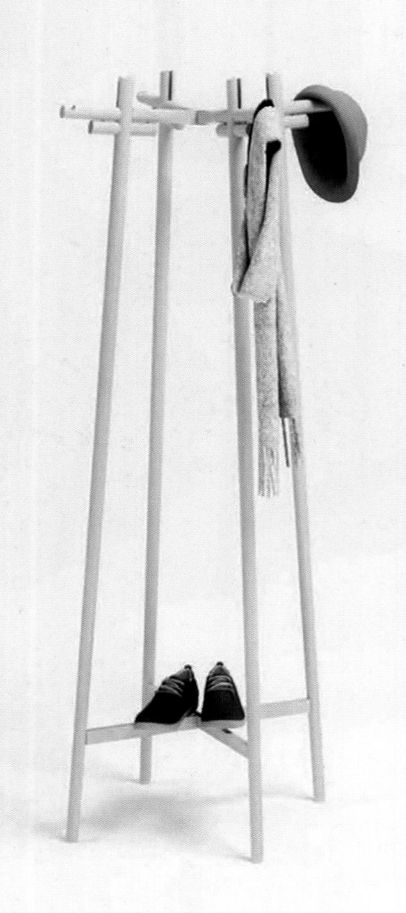

榫卯/扁平化/衣帽架

Tenon / Flat / Cloakstand

材质：榉木/竹

结构：榫卯结构

尺寸：42×36×170cm

重量：7kg

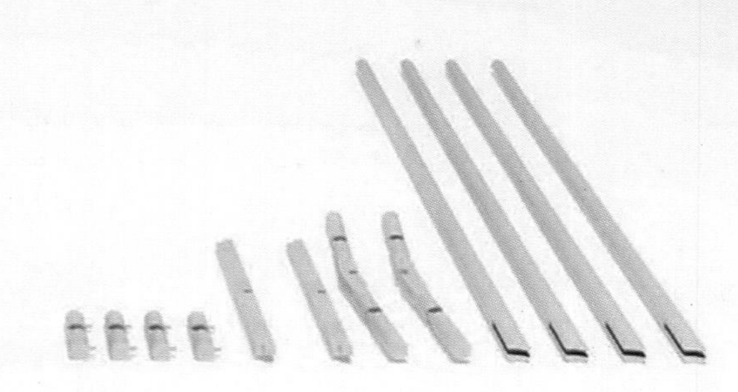

木语
扁平化（榫卯）家具系列
Flat packaging (tenon mortise) furniture series
色彩/束染布 毛毡
素雅/生活美学
榫卯/扁平化结构
工艺/力学收口
功能/实用属性
個性/定义生活语言
① 桌面/Table surface
② 桌腿/Table legs
③ 十字横撑/Cross brace
④ 锁片/locking plate
⑤ 扁平化包装/locking plate

明韵——新中式家具系列
Mingyun— New Chinese Furniture Series

设计师：翟伟民 / Designer : Zhai Weimin

IAI最佳设计大奖

翟伟民 Zhai Weimin

中国建筑学会会员
清华大学文创园区签约设计师
法国创意设计专业委员会会员
南京粽角文化创意设计有限公司创始人
台湾实践大学工业产品设计学系研究所在读

Member of China Architectural Society
Contractual Designer of Tsinghua University Wenchuang Park
Member of French Creative Design Professional Committee
Founder of Nanjing Zongjiao Cultural Creative Design Co., Ltd.
Research Institute of Industrial Product Design Faculty , Taiwan Practical University

本系列休闲家具，汲取了传统木器的榫卯结构，用一种扁平化的设计思路去创作，很好地平衡了家具的材质属性与功能需要以及生活美学之间的关系，拉近了物与人的互动关系，表面的不同材质装饰丰富了产品的视觉语言。扁平化的设计让物，空间与人有了更多的交互可能。

本款椅子设计采用了皮革、毛毡与竹结合，造型风格延续了传统明式家具的造物理念，简洁的线条勾勒出极富有韵味的新中式家具，在结构上严格采用传统的榫卯工艺，在椅子结构的设计上独树一帜，方便折叠，节约了空间。整体比例尺寸符合人体工程学舒适性的需求，在造型的细节处注意线型的渐变和转换，方圆之间以及其他几何形的线型转变，给简洁的造型赋予了新的空间和视角。

This series of leisure furniture draws on the traditional wooden structure and uses a flat design idea to create a balance between the material properties and functional needs of the furniture and the aesthetics of life. The interactive relationship between people and the different materials on the surface enrich the visual language of the product. The flat design allows for more interaction with people and space.
This chair is designed with leather, felt and bamboo combined. The style continues the concept of traditional Ming furniture. The simple lines outline the new Chinese furniture with great charm. The structure is strictly based on the traditional craftsmanship. The design of the chair structure is unique and can be easily superimposed, saving space. The overall proportional size meets the needs of ergonomic comfort. At the detail of the shape, attention is paid to the gradation and transformation of the line shape, and the linear transformation between the squares and other geometric shapes gives a new space and perspective to the simple shape.

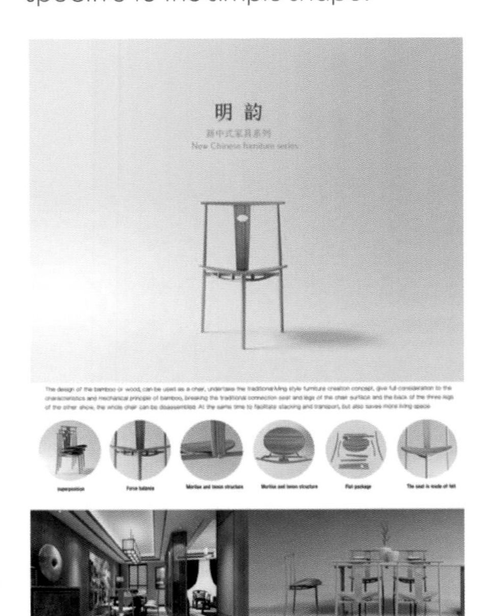

明 韵

新中式家具系列

New Chinese furniture series

明 韵

新中式家具系列

New Chinese furniture series

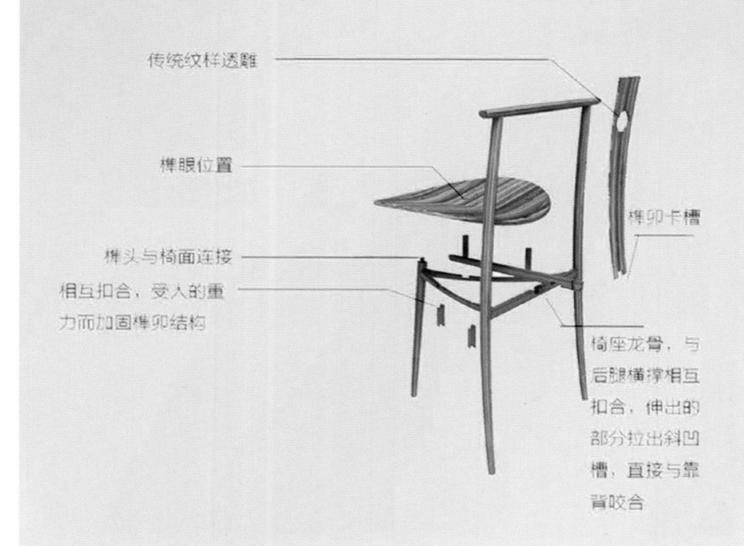

明 韵

新中式家具系列

New Chinese furniture series

superposition

Force balance

Mortise and tenon structure

Mortise and tenon structure

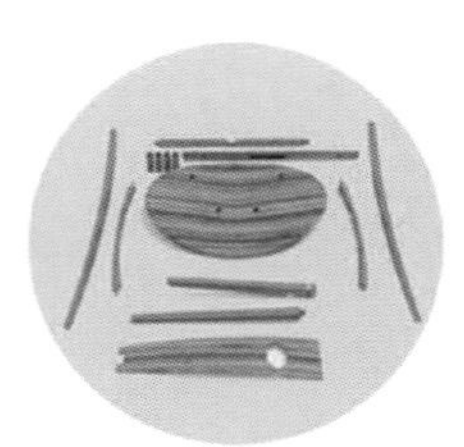

Flat package

The seat is made of felt

高迪系统
Gaudi System

设 计 师：尼科·卡帕 / Designer：Niko Kapa

IAI最佳设计大奖

尼科·卡帕 Niko Kapa

尼科·卡帕是一位屡获殊荣的建筑师、工业设计师和研究员，他是尼克·卡帕工作室的创始人，这是一个支持创新架构和设计的多学科团队，总部设在迪拜。将研究作为设计过程中不可或缺的一部分。他希望发展设计的文化价值，并认识到设计对城市更新、建设社区、教育和职业发展的社会影响。他的工作室项目赢得了众多设计和建筑奖项，赢得了国际认可。

Niko Kapa is an award-winning architect, industrial designer and researcher based in Dubai. He is the founder of Studio Niko Kapa, a multidisciplinary team that supports innovative architecture and design, making research an integral part of the design process. It hopes to develop the cultural value of design and recognize its social impact on urban renewal, community building, education and career development. Studio's projects have won numerous design and architectural awards and have earned international recognition.

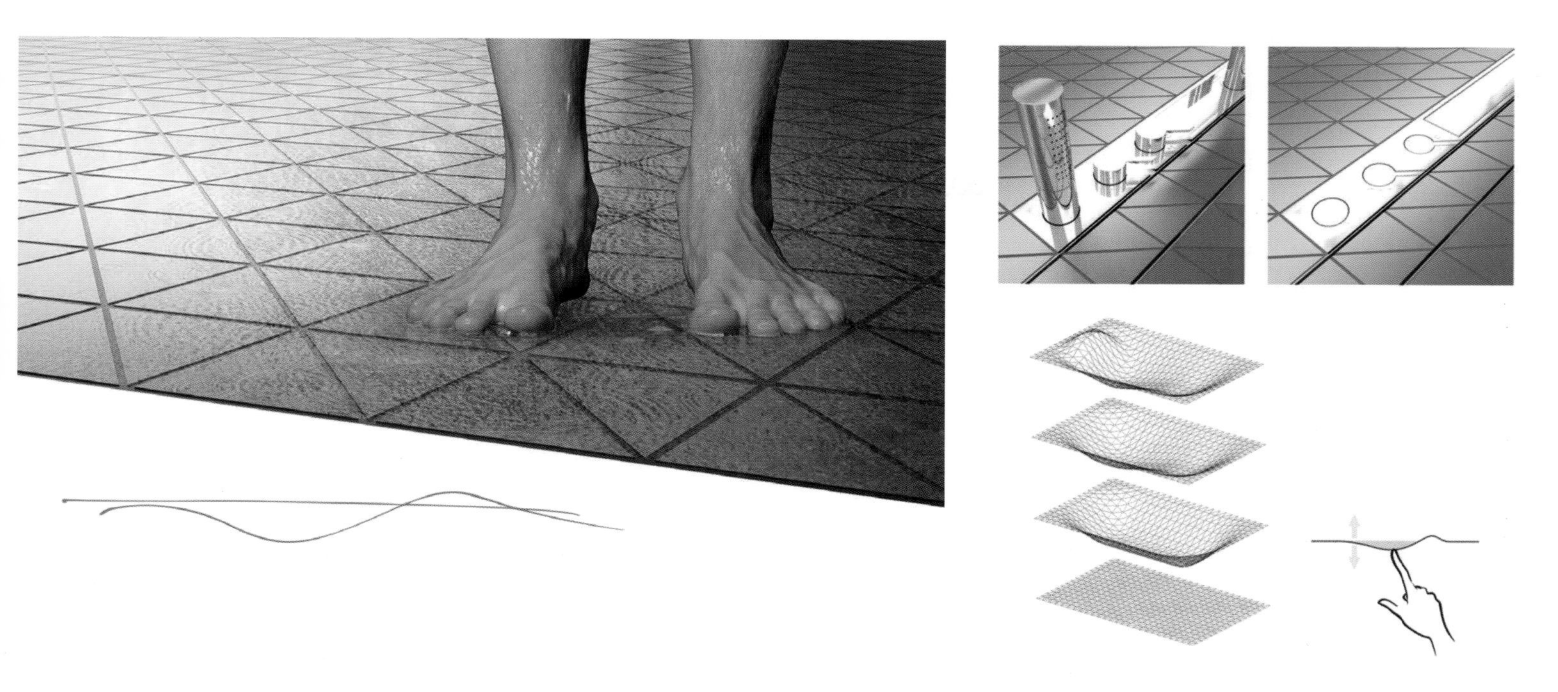

由奥迪创新赞助，高迪系统是一个响应式卫生件，旨在让用户完全控制他们的基本沐浴必需品，“人性化”我们每天使用的对象，同时解决能源效率问题，并通过其设计的关键方面节省资金。考虑到身体姿势和人体工程学，高迪系统可以改变地板并重塑用户决定的沐浴景观。

单个表面通过由抗菌橡胶砖组成的均匀镶嵌图案来组织浴室地板。它们附着在可伸缩的防水聚氨酯涂层氨纶膜上，具有弹性和耐用性。借助完全集成在地板中的液压执行器，可以将表面重新成形为各种形状，以便将淋浴地板转换为人造沐浴景观。隐藏式溢流通道和可伸缩的水龙头可使整个系统与地板齐平，并在不使用时完全消失。

Sponsored by Audi Innovation, Gaudí System is a responsive sanitary piece designed to give users full control of their basic bathing necessities, "humanizing" an object that we use daily, while addressing issues of energy efficiency and subsequently money saving through critical aspects of its design. Taking into account physical posture and ergonomics, Gaudí System can transform the floor and reshape a bathing landscape as decided by the user.
A single surface organizes the bathroom floor through a uniform tesselated pattern that consists of antibacterial rubber tiles. These are attached on a stretchable, water resistant polyurethane-coated Spandex membrane that is resilient and durable. With the help of hydraulic actuators that are fully integrated in the floor, the surface can be reshaped in a variety of formations in order to convert the shower floor to an artificial bathing landscape. A concealed overflow channel and extendable taps allow the whole system to flush with the floor and disappear completely when not in use.

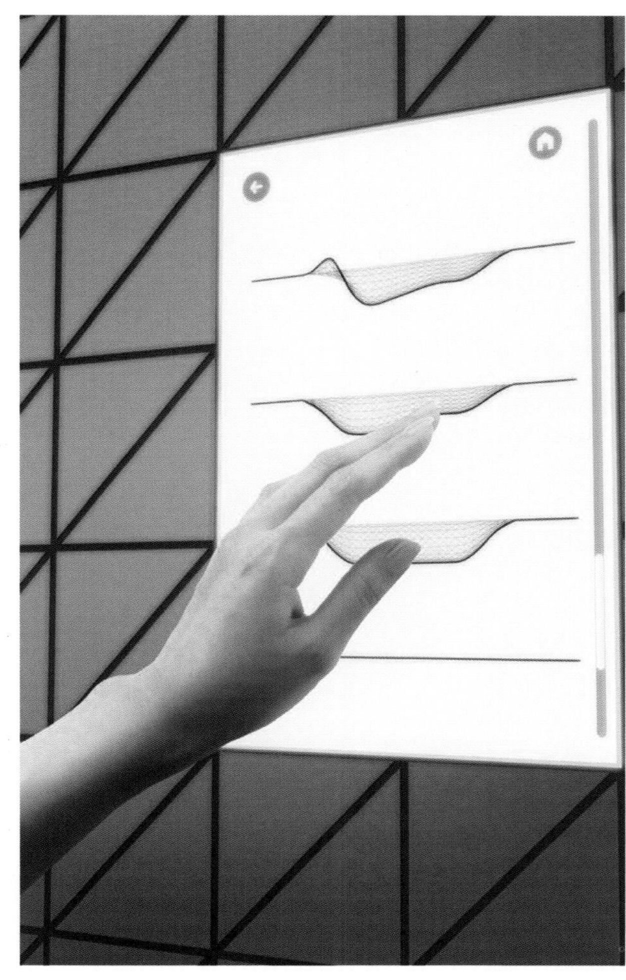

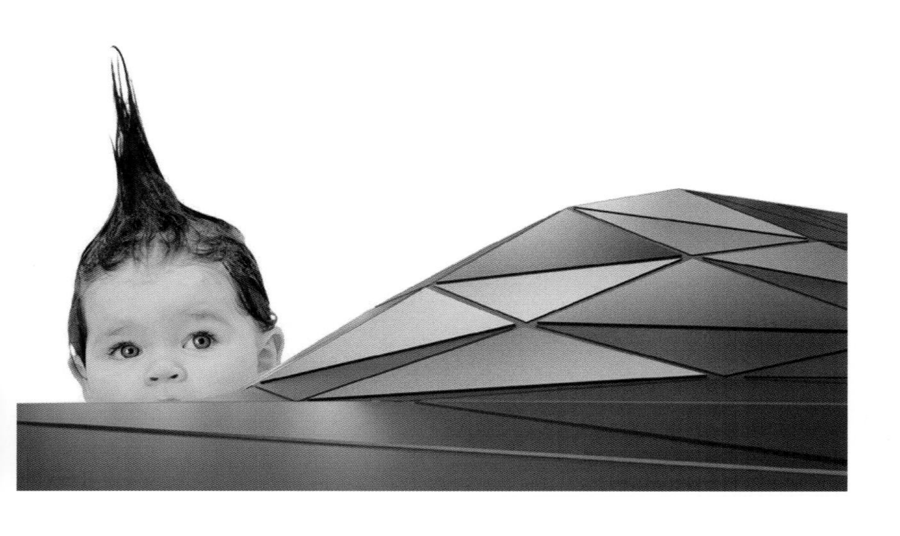

光谱66
Spectrum 66

公司名称：噪咖艺术有限公司 / Organization Name : Noise Kitchen Art Co.,Ltd

IAI 设计优胜奖

噪咖艺术
NoiseKitchen Art Co.,Ltd.

噪咖艺术有限公司为新媒体艺术类策划团队，致力于发展新媒体艺术，将作品结合艺术与科技，并将两个不同领域的概念融入人们的日常生活，创作上习以富视觉刺激的装置呈现与引人入胜的论述观点，平衡艺术与科技的两端，让观者得以在资讯、体验、参观的过程兼获美感与科技性的洗礼，带领民众共享数位艺术生活乐趣。

For the new media art planning team, NoiseKitchen Art Co.,Ltd. dedicated to the development of new media art, combining works with art and technology, and integrating the concepts of two different fields into people's daily life, creating a visually stimulating device with a compelling discussion The idea is to balance the two ends of art and technology, so that the viewer can enjoy the beauty and technology baptism in the process of information, experience and visit, and lead the people to share the joy of digital art life

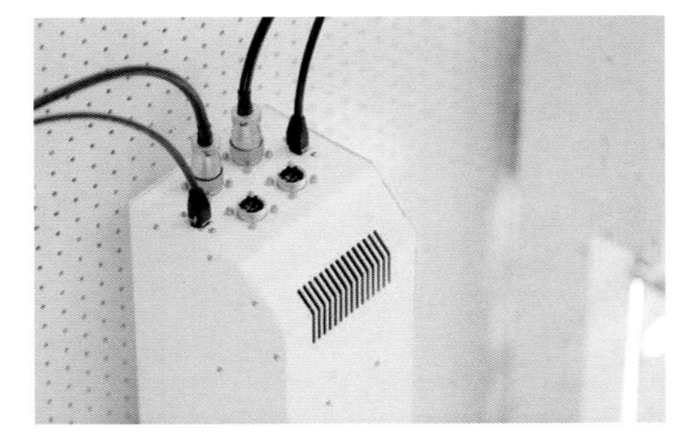

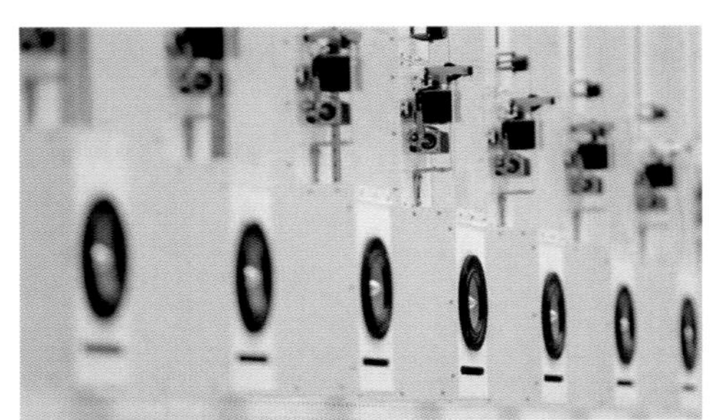

《光谱66》由66组自动机械单弦乐器组成，概念即为可视化的乐谱与曲谱乐音的即时表现。透过高速机械的准确动作，将错落有致的光、音位置以可视化的方式精准演绎。

在此三面包围的空间中，观众有如身处流动的乐谱中，66组乐器随曲谱流动循环往复，《光谱66》作品为上海金桥国际商场空间量身订做，希望观众在此场域舒适的在木质阶梯座席休憩时，亦能聆赏《光谱66》所带来的表演，留下特别又美好的记忆。

Spectrum66 is a custom made spatial installation. It consists of 66 single-string mechanical instruments. The concept of the artwork is to present visualized musical notation and to visually express the sheet music in real-time in a public space.
Surrounded by this three-sided mechanical installation, the audiences are experiencing a space of flowing music. Sounds could appear from different directions as sudden surprises accompanied by the horizontal and vertical movement of lights.
Since Spectrum 66 is permanently installed in a mall, it had to be something exciting also something that relates with its environment. In this space for resting and recreation, we wanted to create work that would grab the visitor' s attention. We also wanted to make something that would last as an extension of a delighted memory.

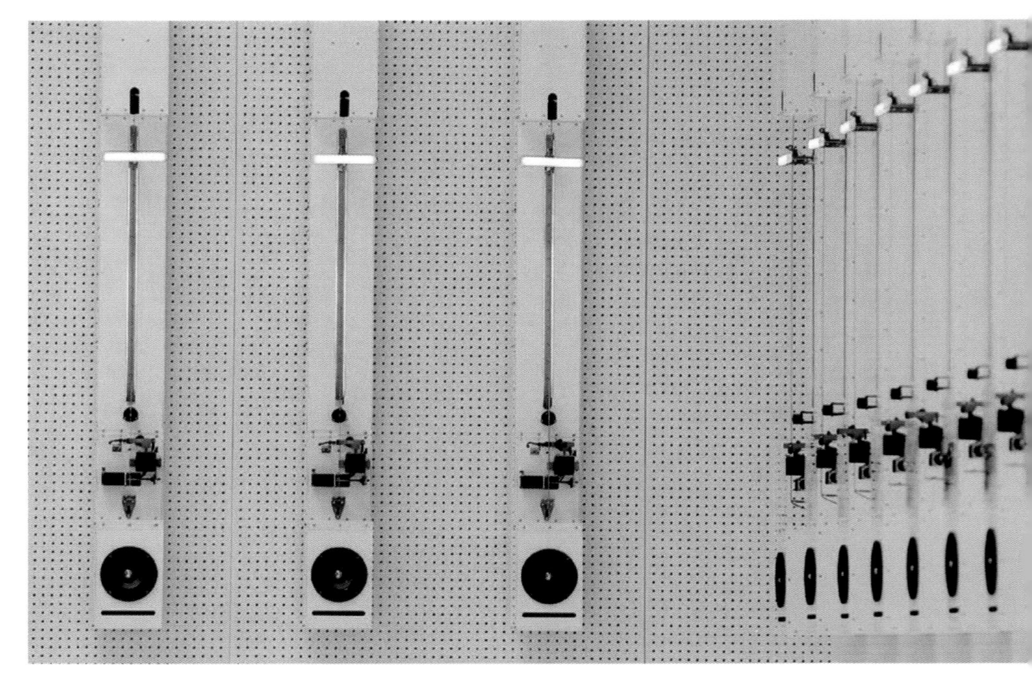

奥迪Cetus
Audi Cetus

设计师：尼科·卡帕 / Designer : Niko Kapa

IAI 设计优胜奖

尼科·卡帕　Niko Kapa

尼科·卡帕是一位屡获殊荣的建筑师、工业设计师和研究员，他是尼克·卡帕工作室的创始人，这是一个支持创新架构和设计的多学科团队，总部设在迪拜。将研究作为设计过程中不可或缺的一部分。他希望发展设计的文化价值，并认识到设计对城市更新、建设社区、教育和职业发展的社会影响。　他的工作室项目赢得了众多设计和建筑奖项，赢得了国际认可。

Niko Kapa is an award-winning architect, industrial designer and researcher based in Dubai. He is the founder of Studio Niko Kapa, a multidisciplinary team that supports innovative architecture and design, making research an integral part of the design process. It hopes to develop the cultural value of design and recognize its social impact on urban renewal, community building, education and career development. Studio's projects have won numerous design and architectural awards and have earned international recognition.

奥迪Cetus是一款氢动力零排放城市车，旨在改变日常城市旅行。我们的想法是为两人创造一辆既可爱又有趣的汽车，努力恢复驾驶体验的刺激。一个有趣的概念练习，真正包含空气动力学，既可以减少能源消耗，也可以成为未来设计美学的一部分。

流畅的曲线和流线型的外壳模仿了鱼的游动的姿态，努力将海洋生物的水动力特性与汽车的空气动力学性能的相似性相结合。这产生了一种充满活力和独特的形式，这种形式由诸如外壳的流动之类的细节构成。

Audi Cetus is a hydrogen-powered, zero-emission city car, designed to change every-day city travel. The idea was to create a car for 2 people that will be likeable and fun, in an effort to restore excitement to the experience of driving. A playful conceptual exercise, truly embraces aerodynamics to both reduce energy consumption and form part of a future design aesthetic.
The smooth curves and streamlined shell mimics the movement of a swimming fish, in an effort to corelate the similarities of the hydrodynamic properties of marine creatures with the aerodynamic performance of the car. This results to an energetic and distinctive form that is formulated by the little details such as the flow of the exterior shell.

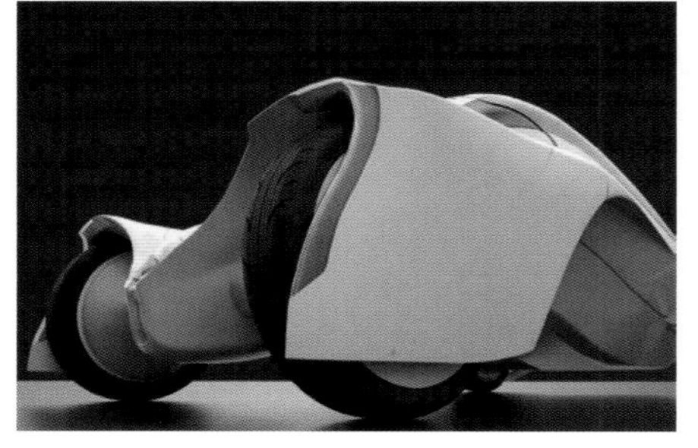

模块化智能LED灯泡
LIGHT 4 LIFE

设 计 师：田角 / Designer : Joe Tian

IAI 设计优胜奖

田角 Joe Tian

工业设计师、交互设计师、工业设计教育者。工业设计硕士毕业于 Savannah College of Art and Design。
无一设计创立人，鲁迅美术学院客座教授，萨凡纳艺术与设计学院助教。作品曾获多项国际大奖并取得专利，部分作品在多国进行展出。

Industrial Designer, Interaction Designer, Industrial design educator.
M.F.A Industrial Design Savannah College of Art and Design,
B.F.A Industrial Design Luxun Academic of fine arts.
Founder of None design Co., Ltd,
Graduate Mentor at Savannah College of Art and Design, Visiting Professor at Luxun Academic of fine arts.
Works have been promoted in multiple international design awards, exhibited in several countries.

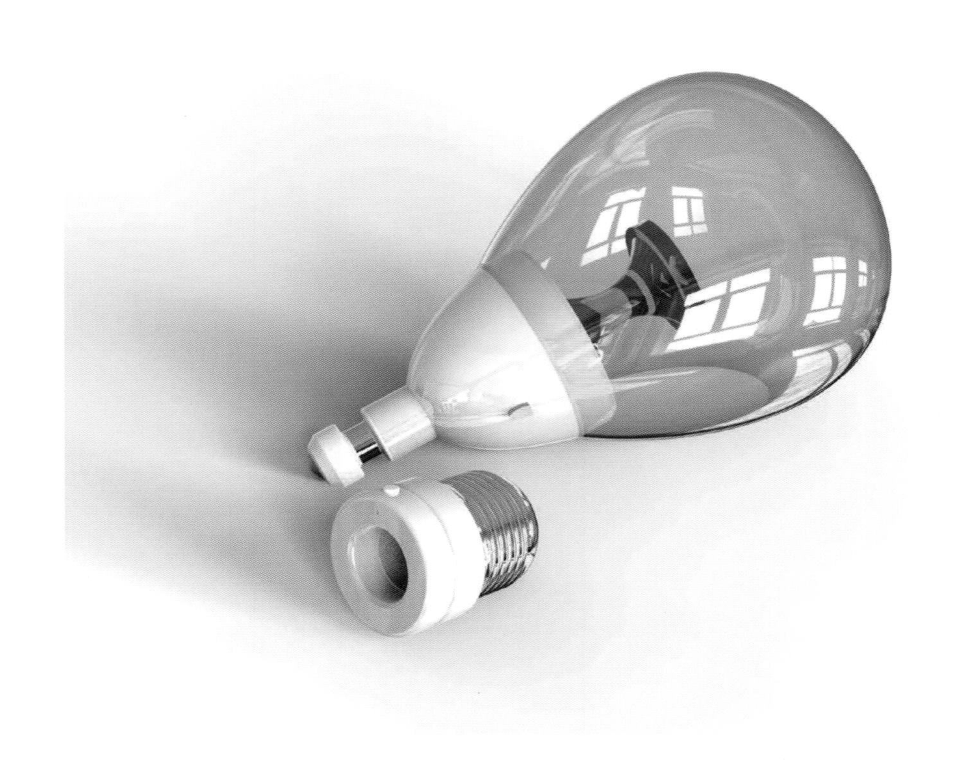

为适应现有接口本设计设计了一个E26接口转换器。降低了废弃灯泡可能会产生的材料浪费问题，使灯泡的每个部分都得以发挥最大的使用寿命。新型接口为行动不便者以及老人更换灯泡提供便利，同时杜绝了更换灯泡时可能会产生的触电危险。经3D模型建立过程中的结构论证，该设计具备量产条件。本设计面向的是家庭用户以及企业用户。该设计预计在面世初期会对传统类型灯泡行业产生一定冲击，灯泡的模块化以及可拓展性将会是吸引消费者的一个创新点。

This design maximized each part's lifetime and give customers more option to choose, the modular design gives this light bulb extensibility. We also designed a conventional socket for the customer who is still using E26 socket. This design decreased wasting problem, maximized lightbulb's lifetime. The new push-push socket gives convenient for elder and people who have difficulty in moving. Based on structure demonstration, this product have massive production possibility. The target users are the household group and enterprise group. This design will impact the light bulb industry. The modular design and expand possibility will be an eye-catching point in the market.

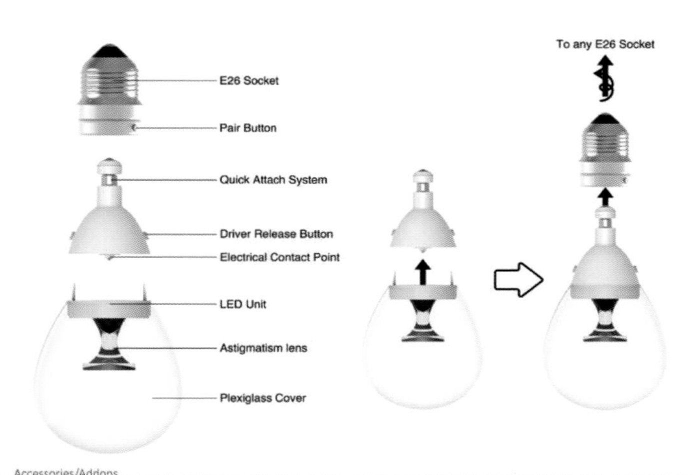

Light 4 Life

Designed by : Jiao Tian, Zhu Wang, Mengnan Sha
In Savannah College of Art and Design

Accessories/Addons

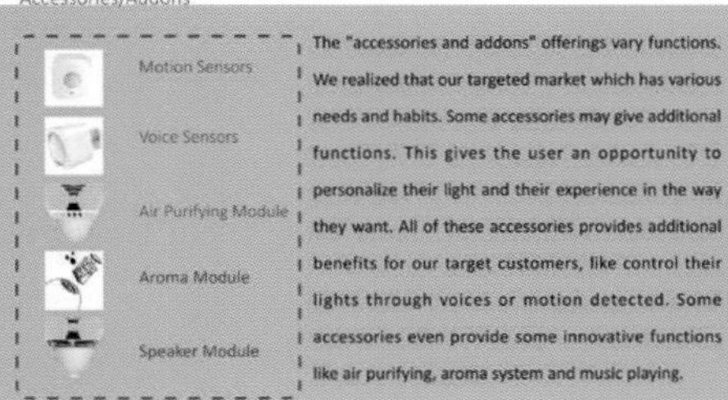

The "accessories and addons" offerings vary functions. We realized that our targeted market which has various needs and habits. Some accessories may give additional functions. This gives the user an opportunity to personalize their light and their experience in the way they want. All of these accessories provides additional benefits for our target customers, like control their lights through voices or motion detected. Some accessories even provide some innovative functions like air purifying, aroma system and music playing.

Control anywhere any device.
Make your life easier.

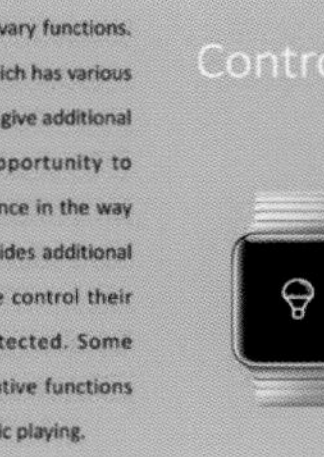

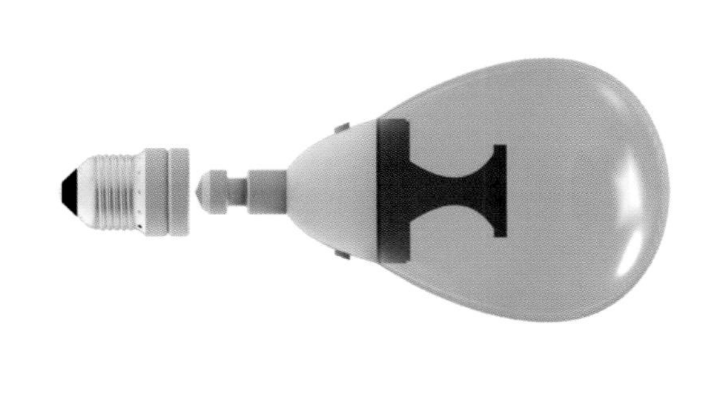

窗户的设计允许尽可能多的光线进入。电致变色玻璃的使用通过按下动力开关来提供调光的可能性，以便控制内部的照明水平。在背面，表面平滑地相互弯曲并包围后轮，形成一个环绕式、可伸缩的行李箱，像抽屉一样打开。其他安全功能包括车辆中的智能传感器。

以柔和的线条和用户友好为特征，表面设计随着流体在道路上运动而流动。在符合人体工程学的试验之后，低调的开发和广泛定制，以最大限度地减少汽车的高度，并随后减少背面的空气湍流。

流线型轮廓的动态进一步强调了流动的自然光滑形式，使汽车具有独特的外观。通过结合这一概念，设计更加直观地表达了空气的平稳流动，这种基本规律赋予了汽车以特色。

The windows are designed to allow as much light as possible to enter. The use of electrochromic glass gives the possibility of dimming by pressing a powered switch in order to control the level of illumination of the interior. On the back, surfaces curve smoothly into one another and embrace the back wheels, creating a wraparound, retractable trunk that opens like a drawer. Additional security features include smart sensors in the vehicle.

Characterized by soft lines and user friendly features, the surface design flows with motion of the fluid around the body that is made to run on the road. The low profile was developed and customised extensively after ergonomic trials in order to minimize the height of the car and subsequently the air turbulence on the back.

The dynamics of the streamlined silhouette are further emphasised by the natural smooth form that flows to give the car a distinctive appearance. By incorporating this concept the design becomes a more visual expression of the smooth flow of air and this underlying order gives the car its character.

DWISS RC1-SW

设计师 / Designer：Rafael Simoes

IAI 设计优胜奖

Rafael Simoes

Rafael Simoes Miranda 自 006 年以来一直致力于设计手表，拥有数百种手表，专为超过 15 个不同品牌和个性设计，包括 Sir 的限量版腕表。

Rafael Simoes Miranda has been designing watches since 2006, with a portfolio of hundreds of watches designed for more than 15 different brands and personalities, including a limited-edition watch for Sir.

DWISS RC1 Automatic是一款带有“神秘”显示系统的手表。小时数通过在多层拨号下运行的光盘表示，并显示12小时的小时数。创造一种神秘的“复杂功能”，而不使用昂贵的技术来改变运动本身。分钟和秒钟以经典的方式用精致的手表表示。

DWISS RC1 Automatic是一款45毫米瑞士制表款，采用316L不锈钢制成。它采用ETA 2824-2高级机芯，配有DWISS“神秘”显示系统。

DWISS RC1 Automatic is a timepiece with our “mysterious” display system. The hours are represented through a disc that runs under the multi-layered dial and shows the hours at 12h. Creating a mysterious"complication", without using expensive techniques to change the movement itself. The minutes and seconds are represented in the classic way with sophisticated hands.

DWISS RC1 Automatic is a 45mm Swiss-made watch fashioned out of 316L stainless steel. It uses ETA 2824-2 topgrade movement, with DWISS “mysterious” display system.

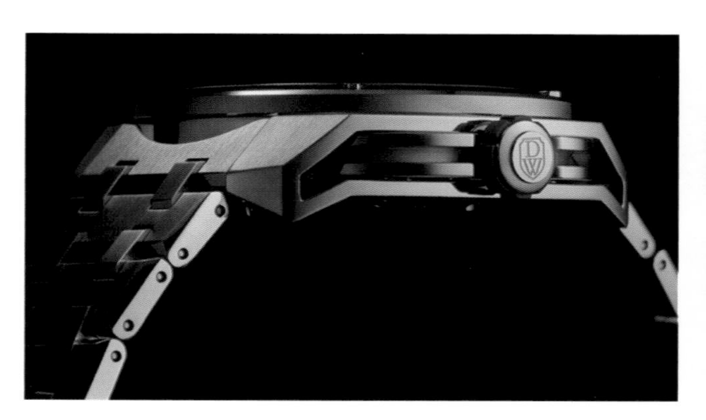

LED 太阳能运动感应灯
LED Solar Motion Sensor Light

公司名称：深圳前海帕拓逊网络技术有限公司 /Organization Name : Patozon

IAI 设计优胜奖

Patozon

前海帕拓逊　Patozon

帕拓逊是国内 A 股上市企业跨境通的全资子公司，是全球消费者美好生活的物质提供商，以"点靓每个人的生活"为使命，为全球个人及家庭提供高品质的消费类产品，客户遍布全球三十多个国家和地区。帕拓逊也是国家认定的"国家高新技术企业"，是集产品设计、研发、销售、服务为一体的高新科技公司。

Patuoxun is a wholly-owned subsidiary of the domestic A-share listed company Cross-border (A-share code: 002640). It is a material provider of a better life for consumers around the world, with the mission of "pointing everyone's life" as the global Individuals and families provide high-quality consumer products with customers in more than 30 countries and regions around the world. Patuoxun is also a national high-tech enterprise recognized by the state. It is a high-tech company integrating product design, research and development, sales and service.

更加环保节能的太阳能感应灯，采用30颗大功率LED灯珠与创新的270度广角设计相结合，光利用率提高30%，照明角度提高50%。一个太阳能灯的照明范围可达215平方英尺，四个可达860平方英尺，灯光穿透力强，即使雨雾天气也能实现照明。SunPower生产的世界领先的太阳能电池板，其光电转换率比其他太阳能电池板高25%。白天吸收太阳能转换为电能储存，在充满电的情况下，在暗光模式中可以工作40小时以上，高亮时光通量达到（200±10）Lm。新开发的PIR热红外人体传感器，其感应范围和灵敏度提高了20%以上，智能区分人与动物所发出的红外波，6秒内感应到人体运动时点亮30秒，如果在照明期间再次检测到运动，它将延长照明时间，感应距离达到39英尺，提供及时的照明需求和有趣的人机交互体验。IP65防水防尘，电路和电池被完全保护，保证了更长的使用寿命。

The more environmentally-friendly and energy-saving solar sensor light uses 30 high-power LED lamp beads combined with the innovative 270 ° wide-angle design to increase light utilization by 30% and increase lighting angle by 50%. A solar light can be illuminated up to 215 square feet and four up to 860 square feet. The light is penetrating and can be illuminated even in rainy and foggy weather. The world's leading solar panels produced by SunPower have a 25% higher photoelectric conversion rate than other solar panels. In the daytime, the solar energy is converted into electrical energy storage. In the case of full charge, it can work for more than 40 hours in the dim mode, and the luminous flux reaches (200 ± 10) Lm in the high light. The newly developed PIR thermal infrared human body sensor has improved sensing range and sensitivity by more than 20%, intelligently distinguishes infrared waves emitted by humans and animals, and illuminates for 30 seconds when sensing human body motion within 6 seconds, if it is detected again during illumination To the sport, it will extend the lighting time to 39 feet, providing timely lighting requirements and an interesting human-computer interaction experience; IP65 is waterproof and dustproof, and the circuit and battery are fully protected for a longer life.

银杏叶椅子
Ginkgo Leaf Chair

设计师：汪宸亦 / Designer : Wang Chenyi

IAI 设计优胜奖

汪宸亦　Wang Chenyi

北京紫金宸室内装饰有限公司总经理、艺术总监
中国建筑装饰协会软装陈设专家委员会专家
“艾舍奖”设计大赛专家评委
中央视电视二台和北京电视台家居节目特约点评嘉宾

Expert of the China Building Decoration Association's Softcovering Beijing Zi Jinchen Interior Decoration Co., Ltd., General manager, art director Expert Committee"Asher award" design competition expert judges Special guest of CCTV 2 and Beijing TV Home Program

这个椅子设计灵感来自中国独有的银杏树叶子。该银杏叶碳纤维一体成型椅子，通过结合碳纤维材料本身具有高强度且重量轻的特点，使得椅子具有抗拉、抗压、抗弯、抗扭、抗剪切、强度高的特点，综合性能更好，解决了稳定性的问题。克服了复杂结构件碳纤维一体成型的技术难点；同时最大限度地降低座椅重量，也解决了椅子腿部设计的力学结构难题。　椅子很轻盈，用户使用方便移动，坚固不变性形，不怕腐蚀，可以同时作为室内和室外的家具，造型优美色彩时尚鲜艳，是居室的艺术点缀。

This chair is inspired by the unique ginkgo leaves from China.This ginkgo-leaf carbon-fibre one-piece chair, combined with the features of high strength and light weight as of carbon fibre material, is enabled with a better overall performance, and with the characteristics of stretching, compressing, bending and shearing resistance and high strength. It overcomes the technical difficulties of one-piece carbon-fibre in complex structure; at the same time minimizes its weight. Moreover, it also solves the mechanical structural problem of the chair leg design. The chair is very light, easy to move, firm and invariant, not easy to corrosion. It can be used as both indoor and outdoor furniture. The sleek colour is fashionable and bright, and can be an artistic embellishment for the living room.

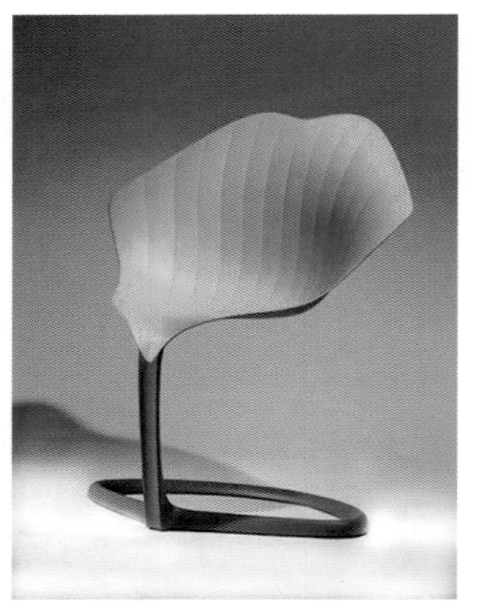

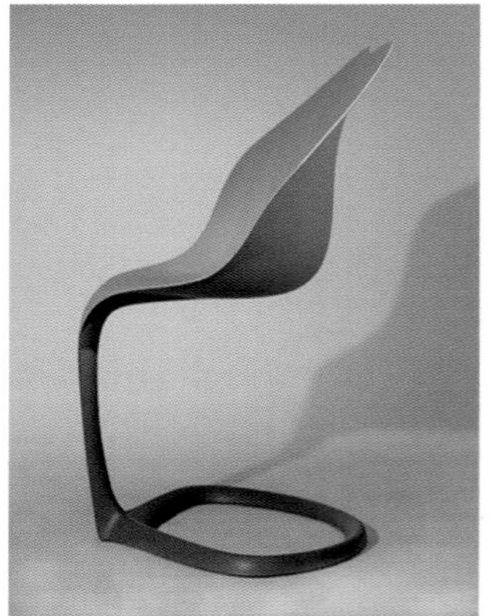

磁吸式无线充电器
Magnetic Wireless Charger

设计师：熊沐国 / Designer : Xiong Muguo

IAI 设计优胜奖

这是一款无线充电器产品，它的主要材料使用铝型材CNC加工一体成型，通过磁吸方式，将金属片与主体吸附形成手机支架，也可以简易地将金属片贴合在主体背面。并设计有镜子的功能，可以作为一个简易化妆镜使用。

This is a wireless charger product. It's main material is formed by CNC machining with aluminum profiles.Metal sheet adsorption main body forming mobile phone holder by magnetic means. It is also easy to attach the metal piece to the back of the main body. It is designed with a mirror function and can be used as a simple make-up mirror.

熊沐国　Xiong Muguo

迄今从业 15 年，发表学术论文若干，设计作品曾获得德国绿色产品设计奖、iF 设计奖、红点奖、美国IDEA产品设计奖、台湾金点设计奖、IAI 设计奖、红星奖、省长杯、醒狮杯等并带领团队获得红点奖、红星奖等国内外设计奖项。

He has been working for 15 years and has published several academic papers. His design works have won the German Green Product Design Award, iF Design Award, Red Dot Award, American IDEA Product Design Award, Taiwan Golden Point Design Award, IAI Design Award, Red Star Award, Governor's Cup. , lion cup and so on and led the team to receive red dot awards, red star awards and other domestic and international design awards.

Communication Design

传播设计

“寻风来”杭州文化品牌形象设计

“Come With Heavenly Wind” Brand Image Design

设计师：冯佳豪 / Designer : Feng Jiahao

IAI 最佳设计大奖

冯佳豪 Feng Jiahao

IAI亚太设计师联盟专业会员、法国创新设计委员会国际设计师会员，2018年毕业于吉林动画学院视觉传达设计专业、曾交流学习于台湾云林科技大学。

A member of the Asia Pacific Designers Federation, an international designer member of the French Innovation Design Council, graduated from JAI Jilin Animation Academy in 2018 with a major in visual communication design. He studied and exchanged at Yuntech Taiwan National Yunlin University of Science and Technology.

“寻风来”杭州文化品牌形象设计
Come with heavenly wind Hangzhou Cultural Brand Design

IDEA | 理念

方言是地域文化的活化石，相较于普通话，方言对于地域文化、事物的表述力道往往更强。
杭州方言虽然适用范围不大，但在东晋时期就已经出现并相传至今，悠久的历史使其独具设计意义。
在全国推广普通话的大环境下，方言文化慢慢在年轻人群中被忽视。据调查，能熟练使用杭州话的年轻人，占比不到9.2%。
设计方案选用了视觉、嗅觉双重感官刺激的独特呈现方式：视觉部分从时下流行的扁平化设计风格入手，辅之以经典杭州词句字体设计、几何平面设计及色彩设计等多维度的设计进行诠释；嗅觉方面则选择了能够代表杭州的独特气味去表现与解读。让更多的人认识、感知杭州地域文化。
通过对原创文化品牌“寻风来”的品牌形象设计全力传达品牌文化内涵，并创造一定的商业价值，正是本次品牌形象设计的意义。

Dialect is the living fossil of the region culture. Comparing with Mandarin,dialect is much stron- ger on the statement of local culture and people.
Although Hangzhou dialect is not very popular, but it has a very long history. As more and more people speak Mandarin, the dialect is being forgot by young people. According to the survey, only 9.2% young people can speak their dialect.
The design scheme selects the unique presentation mode of visual and olfactory dual sensory stimulation: the visual part starts from the popular flat design style, supplemented by the multi-dimensional design of classic Hangzhou font design, geometric graphic design and color design; In terms of smell, he chose to represent and interpret the unique smell of Hangzhou. Let more people know and perceive the regional culture of Hangzhou.
It is the meaning of this brand image design to fully convey the brand culture connotation and create certain commercial value through the brand image design of the original cultural brand “Come with heavenly wind”.

TYPEFACE | 字体

英文字体设计上，结合中文部分，精心设计，体现江南韵味，微风拂动、西湖之水之意。
English font design, combined with the Chinese part, carefully designed to reflect the charm of Jiangnan, the breeze, the meaning of the West Lake.

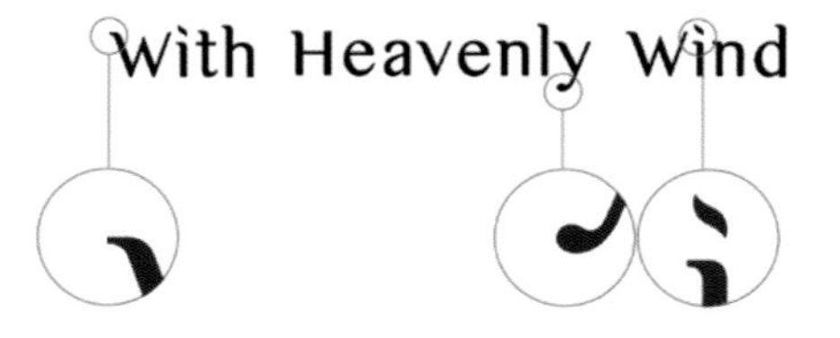

中文字体设计上，既想提取传统山水的韵味及书法的飘逸笔势，以体现杭州江南韵味、风吹柳条、碧波荡漾等寓意，又想保持当代设计的时代性与实用性。因此以标准化的黑体字为结构基础，进行改变设计。

Chinese font design, not only want to extract the charm of traditional landscapes and the ele- gant flow of calligraphy, to reflect the meaning of Hangzhou Jiangnan charm, wind blowing wicker, blue waves and other meanings, but also want to maintain the era of contemporary design and practicality. Therefore, the design is changed based on the standardized blackface.

SLOGAN | 标语

杭州方言 – 声音 – 悠扬回转 – 情怀
杭州茶香 – 气味 – 流动四溢 – 韵味
Hangzhou dialect – sound – melodious turn – feelings
Hangzhou tea fragrance – smell – flowing overflow – charm

方 言 情 怀　茶 香 韵 味

LOGO | 标志

在之前形式推敲的多个稿方案中，比较满意以上理念及方案，最终以这些元素共同构成了项目的标志。

In the previous drafting schemes, the above concepts and schemes are more satisfied, and finally these elements together consti- tute the symbol of the project.

OTHER DESIGN | 其他设计

“来撒倒门”：厉害
“lai sai dao men” means “Awesome”

APPLIED DESIGN | 应用设计

汤沙汇
Tang & Sha

设 计 师：刘珈瑞 / Designer : Liu Jiarui

IAI 最佳设计大奖

刘珈瑞　Liu Jiarui

毕业于中国传媒大学，曾留学巴黎修视觉传达和时尚设计。在传统媒体工作数年，后在本土设计公司任设计总监。近期获得 2016 年伦敦设计节“南京周”海报设计优秀奖和优秀青年新锐设计师称号，伦敦设计节“南京周”年度的设计项目合作伙伴，并赴伦英国参加伦敦设计节“南京周”设计师培训计划。长期合作阿里巴巴、淘宝、森海塞尔、饿了么、惠普、乐视等多品牌的视觉宣传项目。

Jiarui Liu, graduated from Communications University of China, after that ,studied visual communication and fashion design in Paris . After working in traditional media for several years, she worked as a design director at a local design company. Recently won the 2016 Nanjing Design Festival "Nanjing Week" Poster Design Excellence Award and the outstanding young emerging designer title, the London Design Festival "Nanjing Week" annual design project partner, and went to the UK to participate in the London Design Festival "Nanjing Week" design Teacher training program. Long-term cooperation with Alibaba, Taobao, Sennheiser, Ele.me , HP, LeTV and other multi-brand visual publicity projects.

分拆莎士比亚和汤显祖的名字，一是表现两位巨匠的在剧作上的文体书写风格及韵律；二是寓意着东西方文化融合的最好方式是拆开了，揉碎了，再重新组合和书写。即先解构地域性文化，再融合重构成多元性的普世文化。

Split the names of Shakespeare and Tang Xianzu，one is to express the stylistic writing style and rhythm of the two masters in the play, the second is that the best way to integrate the culture of the East and the West is to open it up and smash it. Regroup and write again. That is to first deconstruct the regional culture, and then integrate the universal culture that constitutes pluralism.

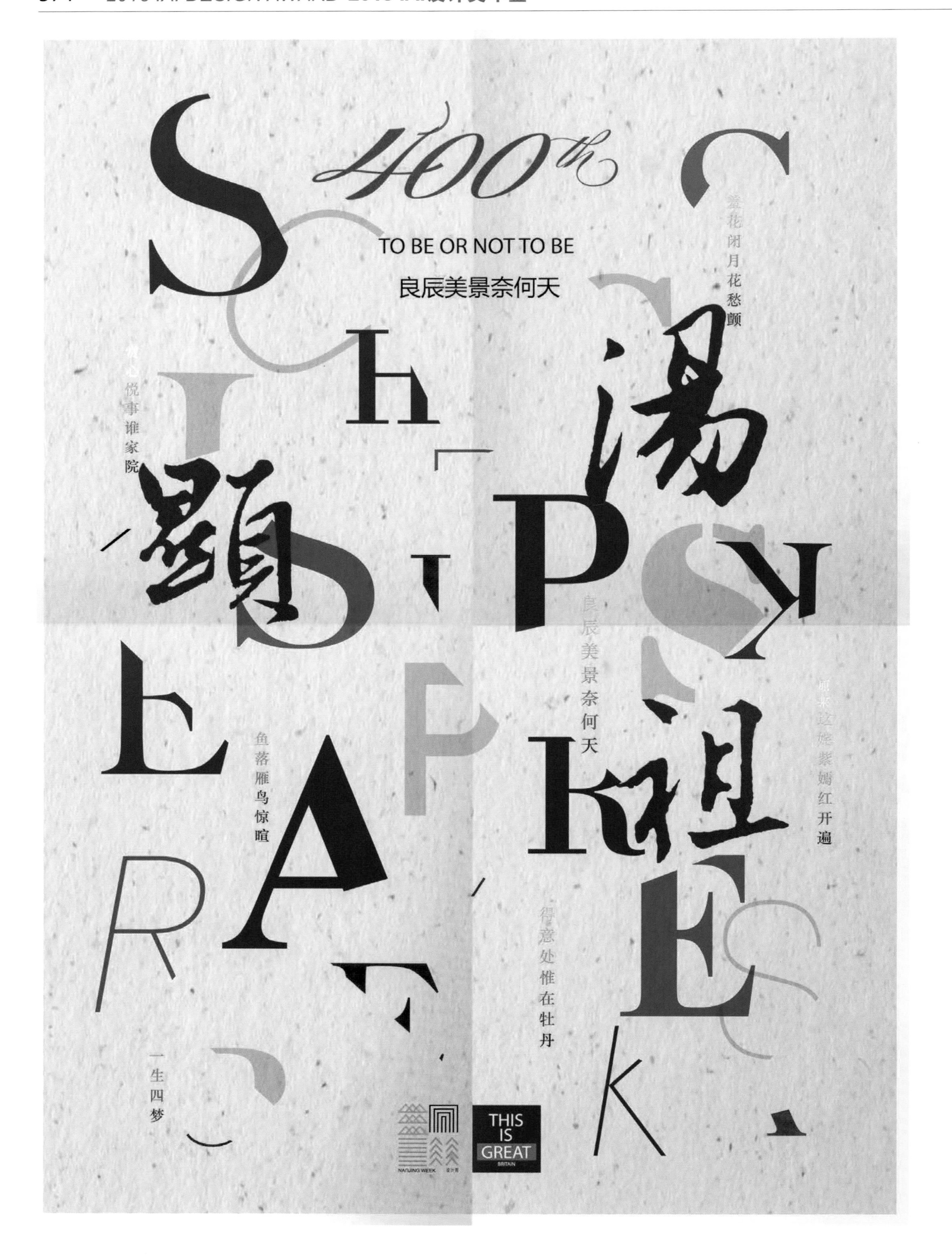
400th
TO BE OR NOT TO BE
良辰美景奈何天
羞花闭月花愁颤
心悦事谁家院
湯顯祖
良辰美景奈何天
原来这姹紫嫣红开遍
鱼落雁鸟惊喧
得意处惟在牡丹
一生四梦
NANJING WEEK
THIS IS GREAT
BRITAIN

汤显祖
Tang & Sha

设 计 师：刘珈瑞 / Designer : Liu Jiarui

IAI 设计优胜奖

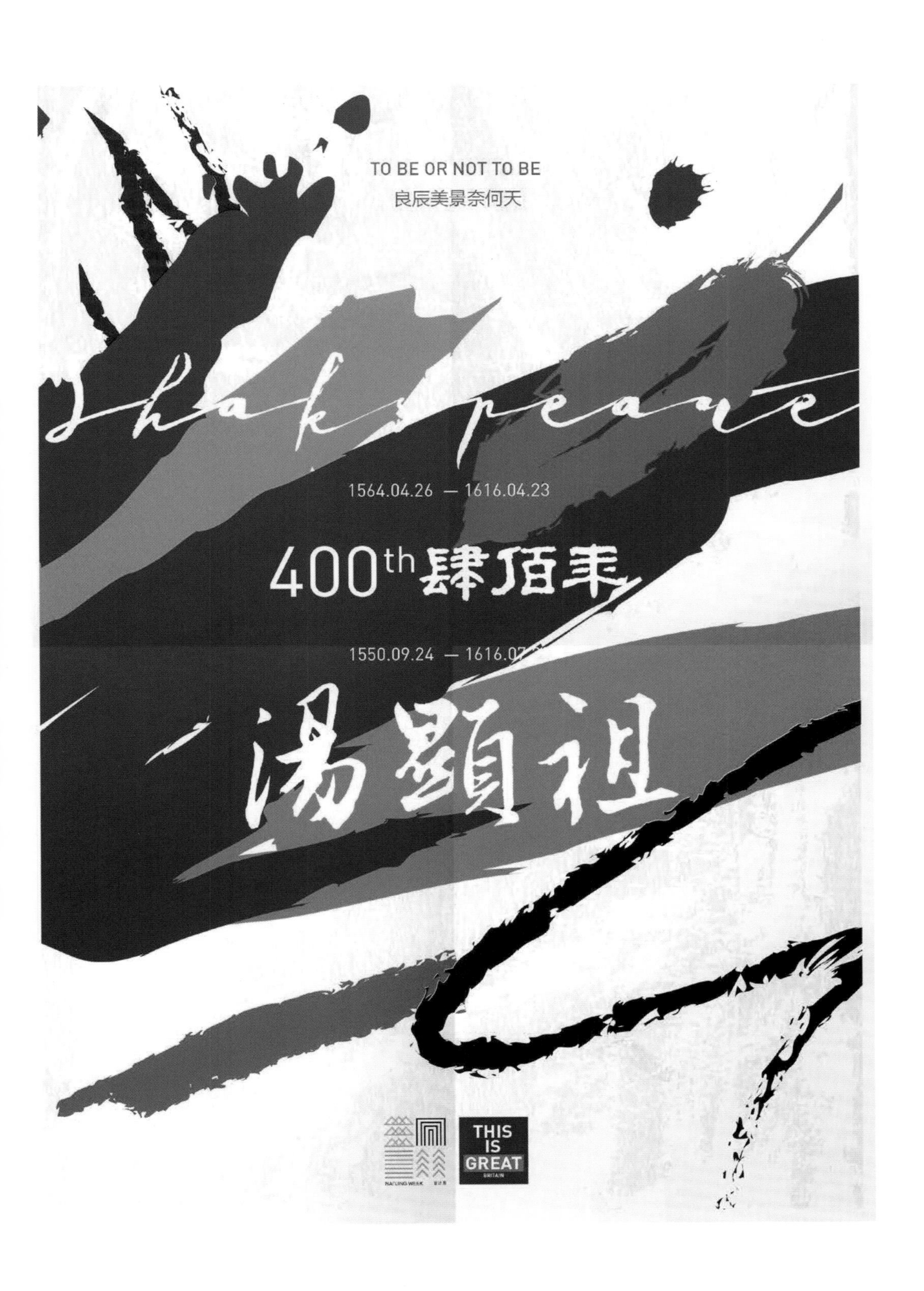

古英文花体字体代表英国文化之传统和经典；傅山的隶书代表中国文化的力度和潇洒，同时傅山是明朝著名文化人，南京是明朝开国之都，有着文化的渊源。红蓝色是英国国旗的颜色，代表英国国粹；淡灰黄底色代表华夏文化源远流长的历史感。曲折的条体现东西方文化冲突和融合的复杂性和必然性。

The ancient English flower fonts represent the traditions and classics of British culture; Fu Shan's Lishu represents the strength and elegance of Chinese culture, while Fu Shan is a famous cultural person of the Ming Dynasty. Nanjing is the founding capital of the Ming Dynasty and has a cultural origin. Red and blue are the colors of the British flag, representing the British national quintessence; the pale gray and yellow background represents the long history of Chinese culture. The twists and turns reflect the complexity and inevitability of cultural conflicts and integration between the East and the West.

明珠酒店品牌设计
Pearl Hotel Brand Design

设 计 师：彭琳 / Designer : Peng Lin

IAI 最佳设计大奖

彭琳 Peng Lin

旗智品牌创始人，专注于成长型品牌国际化创新服务，探寻商业洞察与创意思维的完美契合，用跨界与互联的思维，多维度、全方位的为客户提供极具个性化和具有创造性的视觉沟通设计服务和体验。

Ling Peng, the founder of outofbox brand, focuses on the growing brand, provides innovation service on a global scale and explores the perfect combination between businesas insights and creative thinking. With an all round and multi -dimension perspective, we offer customers personal-ized service and visual experience with the application of internet and cross-boundary cooperation.

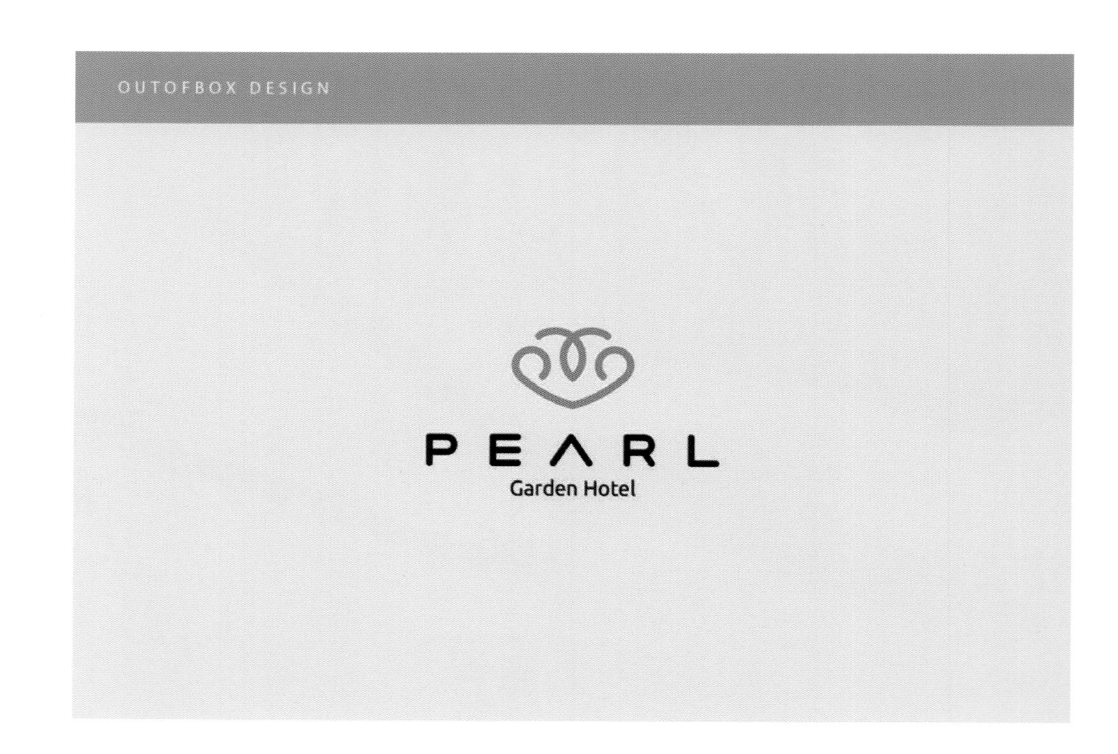

PEARL
PEARL
PEARL

明珠花园酒店
PEARL

明珠花园酒店
PEARL

PEARL
Garden Hotel
PEARL
Garden Hotel
PEARL
Garden Hotel
PEARL
Garden Hotel
PEARL
PEARL

VERY IMPORTANT PERSON
VERY IMPORTANT PERSON
PEARL

PEARL
Garden Hotel
PEARL
Garden Hotel
PEARL
Garden Hotel

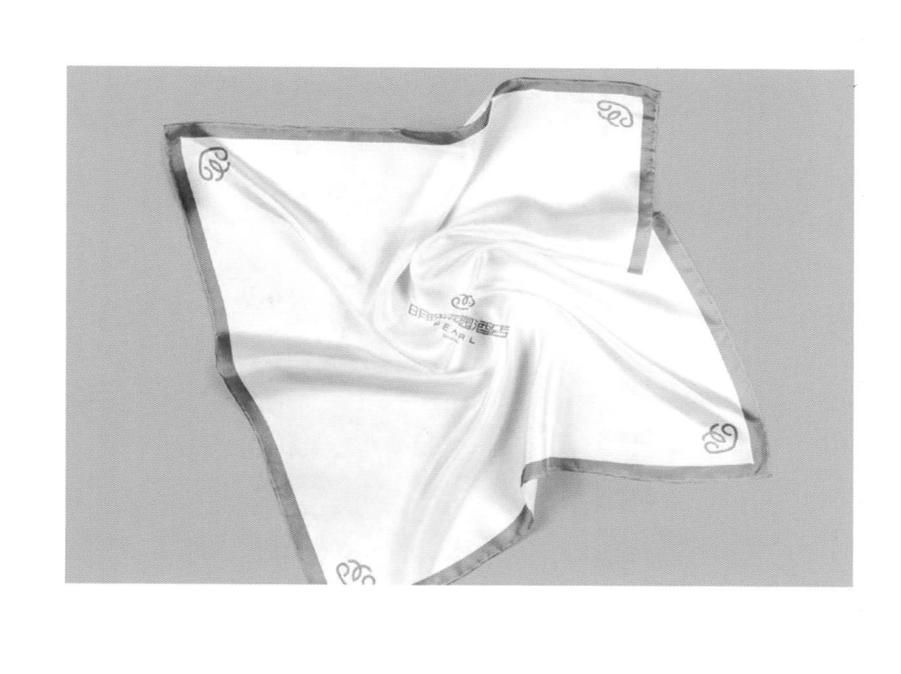

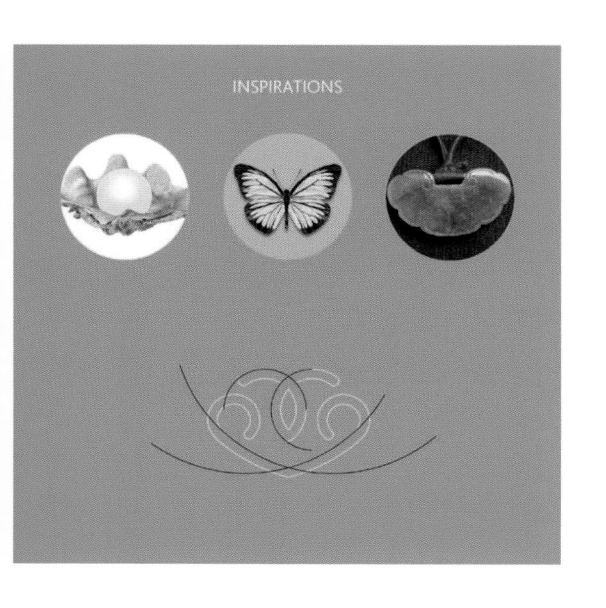
INSPIRATIONS

PEARL
Garden Hotel

常州米朵餐厅品牌设计
Meetto Brand Design

设计师：彭琳 / Designer : Peng Lin

IAI 设计优胜奖

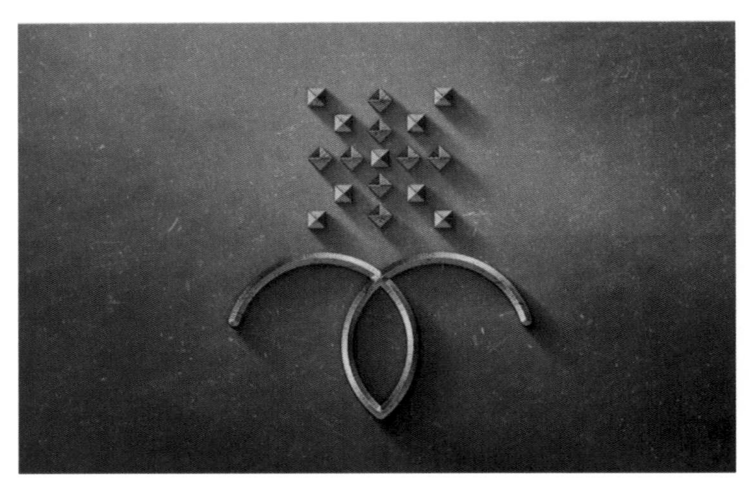

米朵
MEETTO
FOOD&BEVERAGE

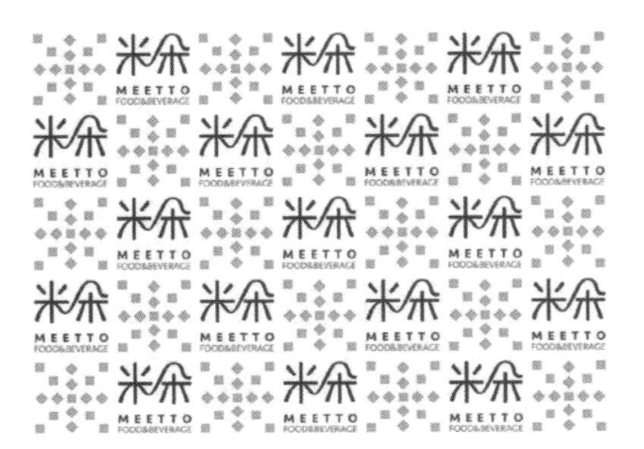
米朵
MEETTO

CAMERICH（锐驰）官方网站

CAMERICH Official Website

公司名称：北京全福凯 / Organization Name : Beijing Quanfukai

IAI 设计优胜奖

CAMERICH 锐驰

锐驰 CAMERICH

作为一个国际化的品牌，锐驰的营销网络已覆盖全球，在英国、美国、德国、澳大利亚、比利时、新加坡等地区开设了 48 家品牌店，在国内 30 多个城市拥有 50 家专卖店，并且这一数字在持续更新。

As an international brand, CAMERICH maintains a global business network, including 48 stores in UK, USA, Germany, Australia, Belgium and Singapore, in addition to 50 franchise outlets in more than 30 cities across China.

锐驰官网涵盖中英文两种语言来满足全球客户的需求。官网在电脑端与手机端各自拥有独立的排版方式，采用响应式布局，能很好地适应各种不同设备的屏幕显示。网站以极简高雅的灰调、动静结合的页面设计完美呈现出品牌动态、产品展示、公司人文情怀三个核心要素，彰显了锐驰兼具国际化与东方美的独特品牌形象。

CAMERICH' s official website was launched with Chinese and English versions to meet different needs of global customers. The website has independent page layouts on PC and mobile phone and enjoys responsive layouts, to adapt to various screen sizes. Combining motion and stillness together, the website uses minimalist, gray-tone page design to display three key elements of CAMERICH: brand dynamics, products details and company culture, revealing a unique brand image that possesses both international and oriental aesthetics.

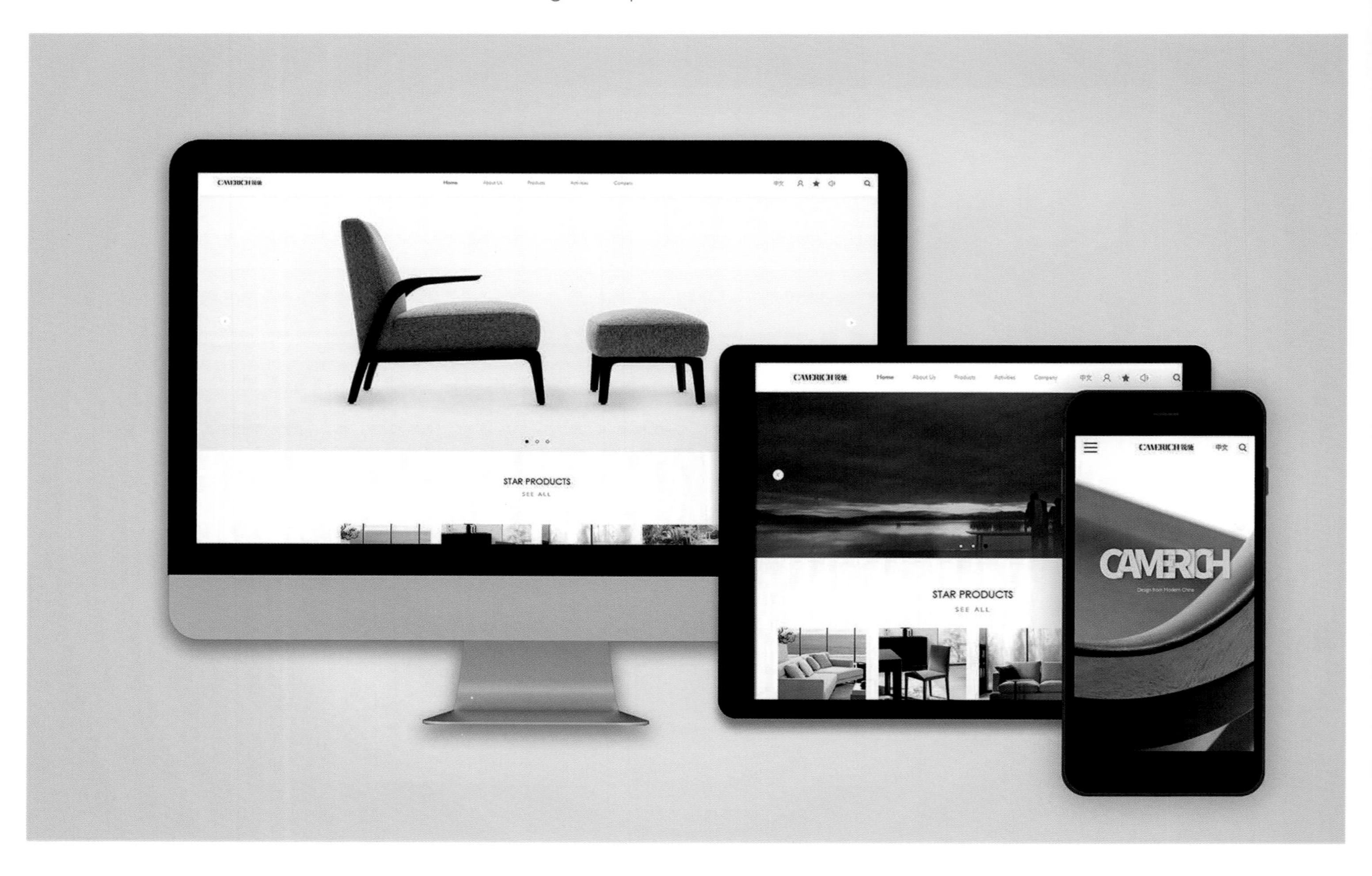

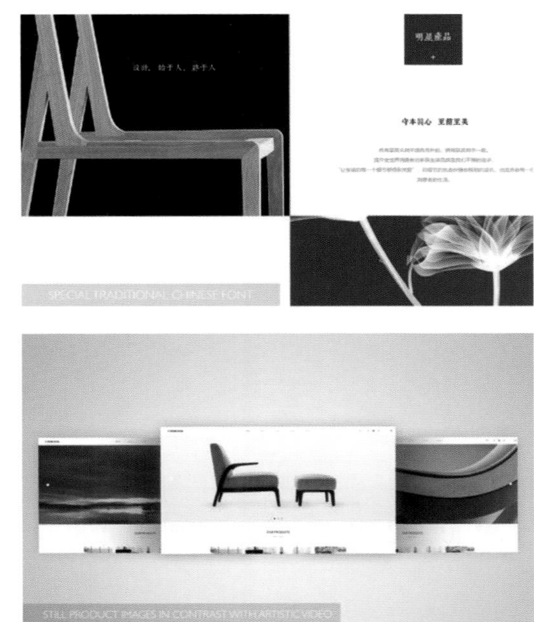
SPECIAL TRADITIONAL CHINESE FONT
STILL PRODUCT IMAGES IN CONTRAST WITH ARTISTIC VIDEO

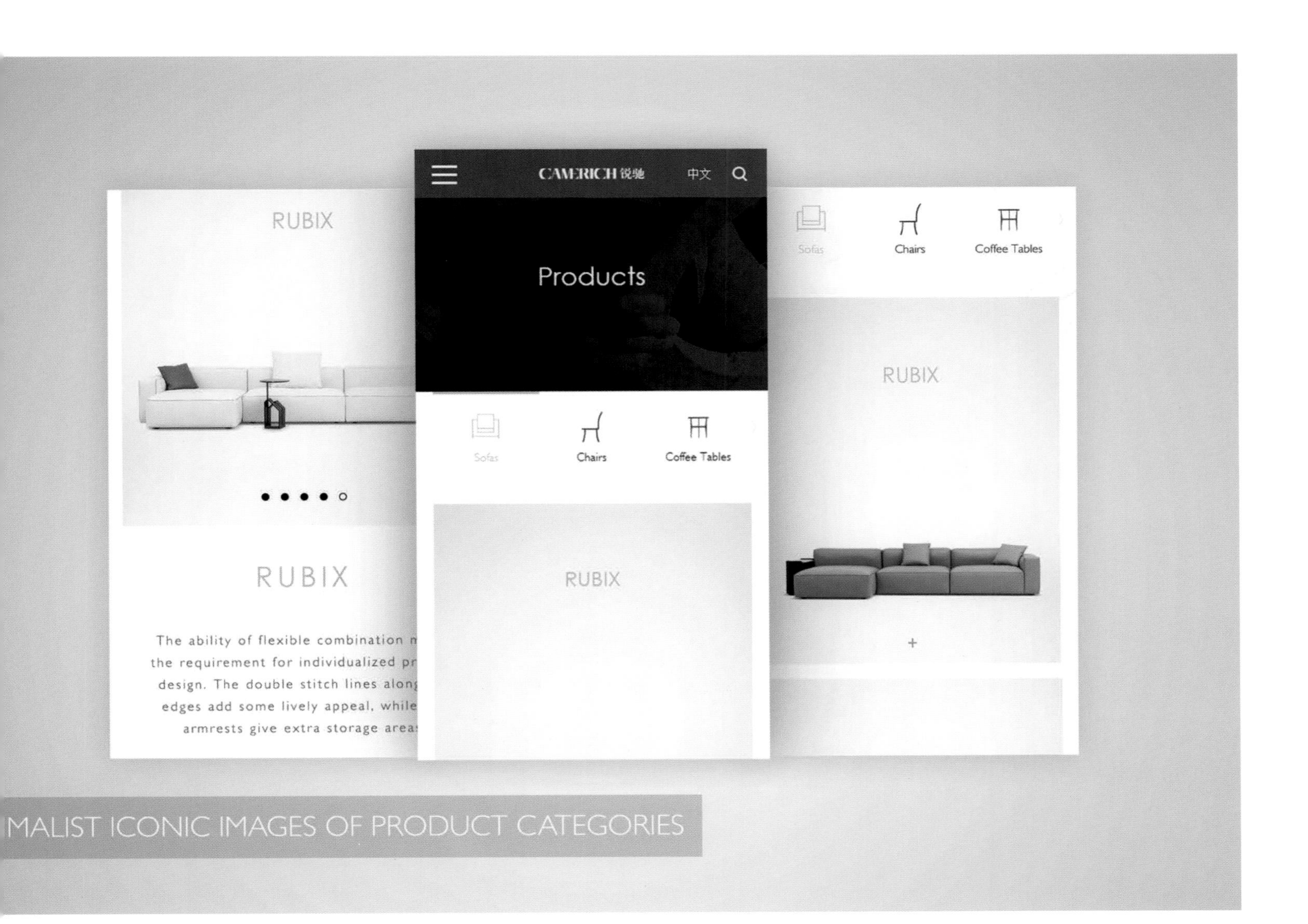
CAMERICH 锐驰
中文
Products
Sofas
Chairs
Coffee Tables
RUBIX
The ability of flexible combination
the requirement for individualized pr
design. The double stitch lines along
edges add some lively appeal, while
armrests give extra storage area
MALIST ICONIC IMAGES OF PRODUCT CATEGORIES

Urban Design and Public Space

城市设计与公共空间

杳无音信
No More Message

设 计 师：胡慧中 / Designer : Hu Huizhong

IAI 设计优胜奖

胡慧中 Hu Huizhong

胡慧中是一位具有建筑背景的艺术家、建筑师和设计师。最近，她获得了由《Perspective》杂志选出在亚太地区40位40岁以下新兴年轻建筑师和设计师的奖项。

Wu Wai Chung is an artist, architect and designer with architecture background. Recently she received 40 under 40 awards of emerging young architects and designers in Asia-Pacific region by Perspective Magazine.

艺术品获得了2018年卢卡双年展的户外艺术家奖。利用Anechoic Room的概念，艺术品呈隧道形状，墙壁特征最大限度地减少了声波的反射。纸张是制作隔音墙的最佳媒体。声音将被最小化并通过隧道改变。当人们走过艺术品时，他们会注意到声音的变化。他们可以通过隧道尖叫或说话，但另一方面声音不会得到如此有效的接收。我们希望传递一条消息，但它永远无法传递。 通过一条无声的道路，出现了混乱。

The Artwork has received the Outdoor Artist Award of Lucca Biennale 2018. Using the concept of Anechoic Room, the artwork is in tunnel shape with wall features that minimize the reflection of sound waves. Paper is the best media to do the acoustic wall. Sound would be minimized and changed through the tunnel. When people walk pass the artwork, they would notice the change of sound. They could scream or talk through the tunnel but sound would not be received so efficiently on the other side. We would like to deliver a message, but it could never be delivered. Through a silent path, chaos arises.

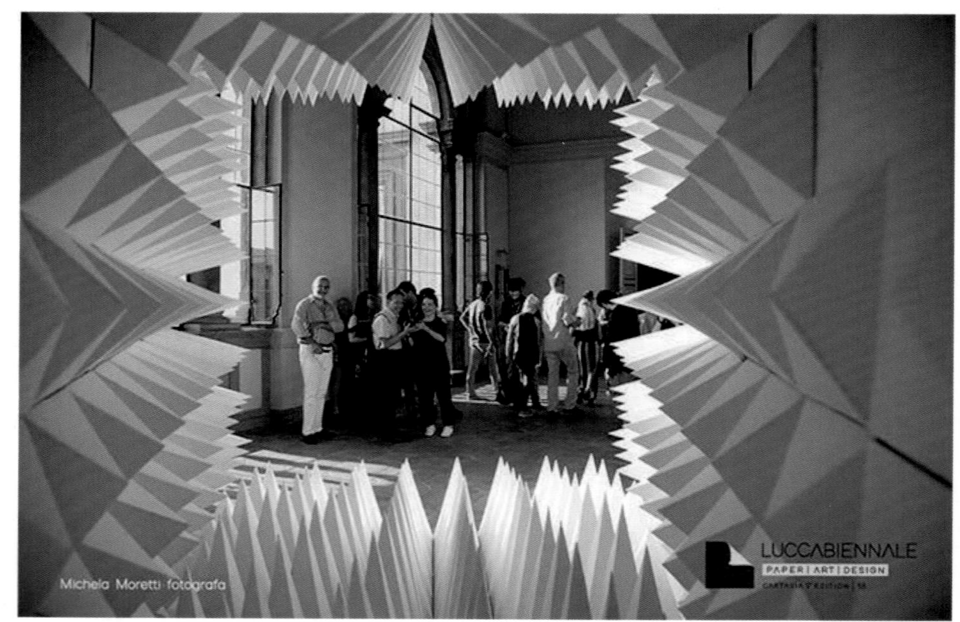

阳光流动
Energy Flow

设 计 师：胡慧中 / Designer : Hu Huizhong

IAI 设计优胜奖

该作品正在通过中国香港政府发展局举办的城市装扮公共艺术比赛获得。它是九龙湾分类的获胜者。当太阳能变成电力时，可再生能源也可以融入艺术品中。在这个城市，我们对环境一无所知，对新鲜事物不再敏感。艺术品由两把摇椅组成，其中一把起身，另一方摇晃到两侧。稍微向后和向前摆动，相互作用。由两把摇椅组合而成的艺术品，配有柔性太阳能板和动力传输USB端口。当他们在椅子上休息时，人们可以给他们的手机充电。该地点位于机电工程署总部的入口处。有一个非常简单的教育路径，专门从事机电工程和可再生能源。因此，这种公共艺术品可以成为教育轨迹的一部分，以提供额外的意义。

The artwork is being realised through winning City Dress-up Public Art Competition organized by Hong Kong Government Development Bureau. It is the winner of the Category Kowloon Bay. When the solar energy turns into electricity, renewable energy can also be integrated into the works of art. In the city, we have become ignorance to our environment, no longer sensitive to fresh things. The artwork is made up of two rocking chairs, one of which gets up and the other party will shake to both sides. Slightly swing backward and forward, interacts with each other. The artwork made by the combination of two rocking chairs, which is equipped with flexible solar sheet with the power transmission USB port. People could charge their mobile phone when they rest on the chairs. The site is at the entrance to the Electrical and Mechanical Services Department Headquarters. There is a very simple educational trail that specializes in electromechanical engineering and renewable energy. Therefore, this public artwork could become part of the educational trails, to provide an additional meaning.

静水流深
Static Water and Deep

设计院校: 北京理工大学 / Designer : Beijing Institute of Technology

IAI 设计优胜奖

北京理工大学
Beijing Institute of Technology

中华人民共和国工业和信息化部直属的一所以理工科为主干，工、理、管、文协调发展的全国重点大学，是国家“211工程”“985工程”首批重点建设高校。

The Ministry of Industry and Information Technology of the People's Republic of China is directly affiliated to the Ministry of Industry and Information Technology. The national key university with coordinated development of work, science, management and literature are the first batch of key construction universities of the national “859 Project” .and “211 Project”

中国传统画作之一的水墨画是国画的代表，也是国画的起源，以笔墨运用的技法基础画成墨水画。从前水墨画以黑白灰为主，水墨画早期都是以山水画的形式来表现的，虽然仅有黑与白，但依然体现着山水风景的层次感与韵律。

现代设计中产品依然可以利用黑白灰体现流动感，《静水流深》以水墨为灵感基础，黑色大理石材质与半透明波纹材质结合，有椅子整体造型的坚毅感与水面的静谧感，又有如瀑布倾泻的动感。

座椅可以放置在博物馆、美术馆或公园喷泉旁，部分博物馆与美术馆整体色调冷淡，如有较暗的环境配合一束强光，就更能体现座椅表面静水与流动对比的美感；街心公园的喷泉旁，座椅的质感与环境完美融合，产品与环境配合，更加给散步的人悠闲自在的体验。

The ink painting of one of the traditional Chinese paintings is the representative of Chinese painting, and it is also the origin of Chinese painting. It is painted in ink based on the technique used in brush and ink. In the past, ink paintings were mainly black and white, and ink paintings were expressed in the form of landscape paintings. Although they were only black and white, they still reflected the layering and rhythm of landscapes.
In modern design, the product can still use the black and white ash to reflect the flow. The "static water and deep" is based on ink. The black marble material is combined with the translucent corrugated material. The firmness of the overall shape of the chair and the calmness of the water surface are as good as The dynamic of the waterfall pouring.
The seats can be placed beside museums, art galleries or park fountains. Some museums and art galleries have a cool overall color. If there is a darker environment with a strong glare, it can better reflect the beauty of the seat surface and the flow of water. Next to the park's fountain, the texture of the seat is perfectly integrated with the environment, and the product is in harmony with the environment, giving the walker a leisurely experience.

时光
Time

设计院校: 北京理工大学 / Designer : Beijing Institute of Technology

IAI 设计优胜奖

《时光》以宇宙的时间感与空间感作为灵感起源，让人们体验时间融入生活、时间流逝及时间自由的理念，我们也在创造一种全新的观看体验，用投影来表现时间，另外产品用到了感应式开关等技术。设计理念在于“让时间休息一会儿”和“将时间洒向空间”，表作为记录时间的载体，当我们需要看时间时它才会展示给我们。光在投射出来的时候缥缈自由，光影与时间，两个至美而又虚幻的媒介，被巧妙地结合到了一起。用太空舱和月球表面形态造型来表现投影表的实体物，产品外观浑圆光滑。“太空舱”的主要舷窗作为投影的镜头，其他两个小舷窗分别为触摸键、提示灯，投射在墙上的指针的灵感来自太空中的群星，时刻与指针像星球转动一般依次跃动出现定格后时刻与分针对应的是表面上的圆形阴影，像宇宙中星球的运转规律。十秒之后，动画效果如倒放一般，又从展开的星球收缩为一个点，最终熄灭。

"Time" takes the sense of time and space of the universe as the inspiration, let people experience the idea of time integration into life, time lapse and time freedom. We are also creating a new viewing experience, using projection to express time, and other products inductive switches and other technologies. The design concept is "to let time rest for a while" and "spread time to space". The table serves as a carrier for recording time. When we need to watch time, it will show us that when light is projected, freedom, light and time Two beautiful and illusory media are cleverly combined. The space capsule and the lunar surface shape are used to represent the physical objects of the projection table, and the appearance of the product is round and smooth. The main porthole of the "space capsule" is used as the projection lens. The other two small portholes are touch keys and reminder lights. The pointers projected on the wall are inspired by the stars in space. The moments and the pointers are like the rotation of the stars. After the freeze, the moments and points are corresponding to the circular shadow on the surface, like the law of the planet in the universe. After ten seconds, the animation effect is reversed, and it shrinks from the unfolded planet to a point and eventually goes out.

第11届IAI全球设计奖获奖名单

Winners List of the 11th IAI Design Award

IAI最佳创新力设计大奖 / IAI Best Innovation Design Award

获奖作品 / WORK	参赛类别 / CATEGORY	获奖者 / WINNER	国籍 / NATIONALITY
(P342) 鳞灯系列	产品设计	戴维德·蒙塔纳罗	意大利

IAI杰出设计大奖 / IAI Outstanding Design Award

获奖作品 / WORK	参赛类别 / CATEGORY	获奖者 / WINNER	国籍 / NATIONALITY
(P016) 层迭	室内设计	张建武、蔡天保	中国大陆
(P160) SLT办公室	室内设计	北京艾奕空间设计有限公司	中国大陆

IAI评审团特别大奖 / IAI Speial Jury Design Award

获奖作品 / WORK	参赛类别 / CATEGORY	获奖者 / WINNER	国籍 / NATIONALITY
(P298) 张家界玻璃桥	建筑设计	渡堂海	以色列
(P342) 鳞灯系列	产品设计	戴维德·蒙塔纳罗	意大利

IAI最佳设计机构大奖 / IAI Best Design Agency Award

获奖作品 / WORK	参赛类别 / CATEGORY	获奖者 / WINNER	国籍 / NATIONALITY
(P320) 哈尔滨菜天下	建筑设计	北京艾奕空间设计有限公司	中国大陆
(P334) 余乐园	空间设计	噪咖艺术有限公司	中国台湾

IAI最佳人文关怀大奖 / IAI Best Humanistic Care Award

获奖作品 / WORK	参赛类别 / CATEGORY	获奖者 / WINNER	国籍 / NATIONALITY
(P260) YM 托儿所	室内设计	大沽·日比野拓	日本
(P252) 我的秘密花园	室内设计	洪逸安	中国台湾
(P012) 一天	室内设计	何丹尼	美国

IAI最佳环境友好大奖 / IAI Best Environmental Friendliness Award

获奖作品 / WORK	参赛类别 / CATEGORY	获奖者 / WINNER	国籍 / NATIONALITY
(P256) CLC北京	室内设计	大沽·日比野拓	日本

IAI最佳设计大奖 / IAI Best Design Award

获奖作品 / WORK	参赛类别 / CATEGORY	获奖者 / WINNER	国籍 / NATIONALITY
(P030) 零宠物俱乐部	室内设计	拿云室内设计有限公司	中国大陆
(P306) 华翔堂社区中心	建筑设计	德科乌维·蒙许	巴西
(P114) X.俱乐部	室内设计	黄永才	中国大陆
(P132) 竞赛机器人实验室	室内设计	Ministry of Design Pte Ltd	新加坡
(P090) 阳朔花梦间酒店	室内设计	姜晓林	中国大陆
(P156) 静界	室内设计	张建武、蔡天保	中国大陆
(P358) 高迪系统	产品设计	尼科·卡帕	希腊
(P140) 春在东方山西展厅	室内设计	朱海博	中国大陆
(P310) 武汉东原售楼处	建筑设计	凌子达	中国台湾
(P016) 层迭	室内设计	张建武、蔡天保	中国大陆
(P328) 小溪家	建筑设计	魏娜	中国大陆
(P324) 倾城客栈	建筑设计	李硕	中国大陆
(P266) HN托儿所	建筑设计	大沽·日比野拓	日本
(P318) 三叠泉	建筑设计	创空间集团	中国台湾
(P078) 洞穴俱乐部	室内设计	黄永才	中国大陆
(P082) 泉上海鲜粥	室内设计	钱银铃	中国大陆
(P252) 我的秘密花园	室内设计	洪逸安	中国台湾
(P020) 水云间茶会所	室内设计	蒋国兴	中国大陆
(P012) 一天	室内设计	何丹尼	美国
(P302) 刚艺红木家具体验馆	建筑设计	黄永才、蔡立东	中国大陆
(P354) 明韵——新中式家具系列	产品设计	翟伟民	中国大陆
(P086) 客语餐厅	室内设计	深圳市艺鼎装饰设计有限公司	中国大陆
(P166) 鼎天办公室	室内设计	黄婷婷	中国大陆
(P024&P136) 茶所、文成堂	室内设计	欧阳昆仑、魏冰	中国大陆
(P373) 汤莎汇	传播设计	刘珈瑞	中国大陆
(P370) 寻风来杭州文化品牌形象设计	传播设计	冯佳豪	中国大陆
(P338) 字由诗	产品设计	易成海	中国大陆
(P376) 明珠酒店品牌设计	传播设计	彭琳	中国大陆
(P348&P367) 磁吸式无线充电器 、智能宠物喂食器	产品设计	熊沐国	中国大陆
(P342) 麟灯系列	产品设计	戴维德·蒙塔纳罗	意大利
(P346) 电子白噪音发生器	产品设计	深圳前海帕拓逊网络技术有限公司	中国大陆
(P347) 机械白噪音发生器	产品设计	深圳前海帕拓逊网络技术有限公司	中国大陆

IAI设计之星 / IAI Design Star

获奖作品 / WORK	参赛类别 / CATEGORY	获奖者 / WINNER	国籍 / NATIONALITY
(P350) 木语——扁平化（榫卯）家具系列	产品设计	翟伟民	中国大陆

图书在版编目(CIP)数据

2018第十一届IAI设计奖年鉴 / 何昌成编. -- 上海 :上海书画出版社, 2019.12

ISBN 978-7-5479-2191-3

Ⅰ. ①2… Ⅱ. ①何… Ⅲ. ①室内装饰设计 – 世界 – 2018 – 年鉴②建筑设计 – 世界 – 2018 – 年鉴③工业产品 – 设计 – 世界 – 2018 – 年鉴 Ⅳ. ①TU238.2-54②TU206-54

中国版本图书馆CIP数据核字(2019)第205330号

2018第十一届IAI设计奖年鉴

何昌成 编

责任编辑 吴蔚 吴雪莲
技术编辑 钱勤毅
装帧设计 何昌成

出版发行 上海世纪出版集团
上海书画出版社

地址 上海市延安西路593路 200050
网址 www.ewen.co
www.shshuhua.com
E-mail shcpph@163.com
印刷 上海普顺印刷包装有限公司
经销 各地新华书店
开本 635×965 1/8
印张 49 字数 200千字
版次 2020年1月第1版 2020年1月第一次印刷

书号 ISBN 978-7-5479-2191-3
定价 380.00元

若有印刷、装订质量问题，请于承印厂联系